Superconductivity: Discoveries and Discoverers

Kristian Fossheim

Superconductivity: Discoveries and Discoverers

Ten Physics Nobel Laureates
Tell Their Story

 Springer

Kristian Fossheim
Department of Physics
Norwegian University of Science and
 Technology
Trondheim, Norway

The author of this book has received financial support from the Norwegian Non-fiction Literature Fund.

ISBN 978-3-642-36058-9   ISBN 978-3-642-36059-6 (eBook)
DOI 10.1007/978-3-642-36059-6
Springer Heidelberg New York Dordrecht London

Library of Congress Control Number: 2013937316

© Springer-Verlag Berlin Heidelberg 2013

This work is subject to copyright. All rights are reserved by the Publisher, whether the whole or part of the material is concerned, specifically the rights of translation, reprinting, reuse of illustrations, recitation, broadcasting, reproduction on microfilms or in any other physical way, and transmission or information storage and retrieval, electronic adaptation, computer software, or by similar or dissimilar methodology now known or hereafter developed. Exempted from this legal reservation are brief excerpts in connection with reviews or scholarly analysis or material supplied specifically for the purpose of being entered and executed on a computer system, for exclusive use by the purchaser of the work. Duplication of this publication or parts thereof is permitted only under the provisions of the Copyright Law of the Publisher's location, in its current version, and permission for use must always be obtained from Springer. Permissions for use may be obtained through RightsLink at the Copyright Clearance Center. Violations are liable to prosecution under the respective Copyright Law.

The use of general descriptive names, registered names, trademarks, service marks, etc. in this publication does not imply, even in the absence of a specific statement, that such names are exempt from the relevant protective laws and regulations and therefore free for general use.

While the advice and information in this book are believed to be true and accurate at the date of publication, neither the authors nor the editors nor the publisher can accept any legal responsibility for any errors or omissions that may be made. The publisher makes no warranty, express or implied, with respect to the material contained herein.

Printed on acid-free paper

Springer is part of Springer Science+Business Media (www.springer.com)

Preface

This book has a dual perspective; on one hand the often underexposed human side of the life of outstanding scientists, on the other hand the hard facts about how great scientific achievements were made. I would like first to explain this perspective.

Humanity has always benefited greatly from courageous forerunners of progress, new knowledge and useful insights. Some of them have possessed true genius. But what are the circumstances and processes behind great scientific discoveries and profound intellectual advances? What are the driving forces and the motivating push? What roles do upbringing and family background play? Are there some little illuminated common denominators? In what way are the forerunners different from most people? Or, are they after all not so different, except being exposed to experiences that shape them and provide circumstances and opportunities that promote and stimulate their talents more than usual? How can we find out?

In the processes towards the award of the Nobel prizes in natural science a systematic effort is made to find and honour the forerunners, the discoverers of profoundly new truths, those spearheading the quest to move on into new knowledge territory. Surely, documenting 10 profiles and reminiscences like we do in the book you are reading here, will not provide the ultimate answer. Yet, I am convinced there is something to learn from these stories. On the one side there was a great diversity of conditions under which scientific breakthroughs happened, on the other hand there are common features. It is up to the reader to discover the possible red thread.

The second, and main objective of the book is its scientific subject, superconductivity. This is a field where I greatly enjoyed confronting 10 Nobel laureates one at a time and unravel their stories. This field has the advantage of 100 years of continuous development. Thousands of scientists have worked there and had their own dreams, a dozen of them becoming Nobel laureates, from Heike Kamerlingh Onnes in 1913 to Abrikosov and Ginzburg in 2003.

The basis for this book was first laid in 2001 when I completed a round of videotaped interviews of seven Nobel laureates, each of 1–2 hours length, later to be supplied with two more dialogues in 2003, and yet another one in 2004. The topic, superconductivity, is very much a personal, lifelong fascination since I started research in that field during two years at University of Maryland, 1965–67.

The material was originally intended for, and used in some highly compressed, brief biographical notes in a special chapter of a book I was writing with my physics colleague Asle Sudbø at the Department of Physics at the Norwegian University of Science and Technology in Trondheim. The book was published with the title *Superconductivity. Physics and Applications* (Wiley 2004).

However, having all of this quite original historical material at hand, and doing nothing more about it, did not seem right. Possibly, donating it to an internationally renowned archive, like the Bohr Archives in Copenhagen, could be an alternative, but not fully satisfactory from the point of view of accessibility, and surely also not from an editorial perspective.

Several years after the interviews took place, in 2010, I approached Springer and asked if they were interested in publishing the material, and indeed they were. They had two specific requirements: Each interview should be accompanied by a brief biography, and all questions from the interviewer (KF) should be removed. I was sceptical, but when I tried it out, it did not compromise the contents to any significant extent. The book is unique in the sense that it follows a central theme of physics during as much as sixty years through the stories of 10 Nobel laureates, as told by each of them. It is hoped that it may provide inspiration to new generations of physicists, and even reach a wider audience.

My tasks during recent years as Vice President, and later as President of The Royal Norwegian Society of Sciences and Letters made the work a slow process, so we missed printing the book during the centennial of superconductivity in 2011. Nevertheless, here it is.

Superconductivity covers a huge span of ideas and applications. It takes you from the deepest science, like the much heralded Higgs mechanism, to the most fantastic technology, like recording of magnetic signals generated in your brain by your thoughts and allowing trains to be levitated by superconductivity while travelling at a speed of more than 500 km/h. It turns out that there is no natural phenomenon better suited for the study of the Higgs mechanism than superconductivity. The evidence dates back to 1933, with the discovery of the Meissner effect! Moreover, the experimental facilities, leading to the recent disclosures in the CERN laboratories in Geneva of the likely observation of the Higgs particle, are themselves the greatest manmade scientific wonder in human history. The whole experiment would be totally impossible without the use of superconductivity technology. Due to the recent great interest in the Higgs mechanism, I invited my theory colleague Asle Sudbø to write a special chapter on the Higgs mechanism in superconductivity. It may come as a surprise, even to many condensed matter physicists that the widely known and used theory for superconductivity created by Ginzburg and Landau in 1950, provides an excellent basis for the description and understanding of the Higgs mechanism.

I have to acknowledge a lot of people for their kind assistance. Foremost among these, naturally, are the 10 Nobel Laureates who, without exception, agreed to meet me and dig into their own personal and scientific history, and for most of them, with amazing accuracy. Most interviews were conducted in the laureates' offices, or in their home. In Europe: in Zurich, Paris, and Cambridge. On the East Coast of the US: in Providence, Schenectady, Princeton, and Tallahassee. On December 11, 2003, the day after receiving their prizes, Abrikosov and Ginzburg were separately interviewed in Grand Hotel, Stockholm, where I found that even the oldest one of the interviewed physicists, Vitaly Ginzburg, at 87, was still sharp as a needle. Since then, two of these 10 great men, Ginzburg and de Gennes passed away. I feel privileged to have captured their impressive stories and parts of their unique personalities through their own words. It seems the time was ripe. Finally, in 2004 I interviewed Josephson in Cambridge.

Rather than polishing the language, I have kept the informal, oral, sometimes lively style from the interview situations in full understanding with Springer. To finalize all interview material, I invited all laureates who are still among us, to read and correct their own interview. In the case of de Gennes, I was kindly assisted by his wife Françoise Brochard-Wytt and his former colleague Etienne Gyon. At the National High Magnetic Field Lab in Tallahassee, the director, Greg Boebinger, was very helpful. Schrieffer had already given me his full consent to use the interview in whatever way I wanted during our interview meeting. I take personal responsibility for the final version of the Josephson text, written as a brief summary in third person since he did not participate in the final round of reading. I also found support in Josephson's Nobel lecture available on Internet. In the presentation of Vitaly Ginzburg, I take the responsibility for the interpretation of the tapes since he was not among us any more when I finalized the transcripts. The soundtrack was clear enough. In all other cases the laureates had every chance to make corrections.

People have commented on my choice of 10 laureates, that it could have been different. My choice was based on the following: First of all, the topic was from the start limited to superconductivity. Secondly, since the book is based on interviews, only those who were still among us during the interview period 2001–2004 could be included. Two of them, de Gennes and Anderson, were not specifically awarded the Nobel Prize for their work in superconductivity. However, it is my personal judgement that their work deeply stimulated the science of superconductivity, and hence should be included. I could have included work on superfluids as well, but chose not to, mostly just to limit the whole project.

The brief biographical notes introducing each interview are just short sketches, or summaries of what I thought were the most interesting and relevant aspects of their story. These are based on the interviews and on available literature, in some cases CV's and personal biographies, in other cases on experiences from scientific collaboration, like with Bednorz and Müller, or on my previous extended biographical work, like in the case of Giaever. To the extent that the biographical notes are similar to those published in the Wiley textbook in 2004, I am indebted to Wiley for permitting me to use those texts or parts thereof whenever I wanted.

Finally, I should like to express my special appreciation for the encouragement received from Claus Ascheron at Springer. Without his continuous support, this book would probably not have materialized. Similarly, I would like to express my gratitude to the Norwegian Non-fiction Literature Fund for a stipend which was, most of all, a mental stimulus confirming to me that the project was worthwhile, also from a broader perspective of society.

Trondheim, Norway
October 2012

Kristian Fossheim

Contents

1	**Introduction**		1
2	**Vitaly L. Ginzburg: The Ginzburg-Landau Theory of Superconductivity**		**9**
	2.1	Biographical Notes	9
	2.2	His Own Story	10
		2.2.1 Early Years	10
		2.2.2 Education	11
		2.2.3 War Years, Tamm, and Sakharov	11
		2.2.4 Superconductivity, Landau	12
		2.2.5 Superfluids and Low Dimensional Superconductors	13
		2.2.6 How GL Theory Came to Be	14
		2.2.7 Remarks About the Prize	16
3	**Alexei A. Abrikosov: The Magnetic Structure of Type II Superconductors**		**17**
	3.1	Biographical Notes	17
	3.2	His Own Story	18
		3.2.1 Childhood in Moscow	18
		3.2.2 School and Education	19
		3.2.3 Entering the Landau Group	20
		3.2.4 With Landau	21
		3.2.5 Superconductivity: Discovery of Type II	22
		3.2.6 Difficulties with Landau	26
		3.2.7 Acceptance	26
		3.2.8 Side Issues, the Bomb, KGB, the Prize	27
4	**Leon N. Cooper: The Microscopic Theory of Superconductivity**		**29**
	4.1	Biographical Notes	29
	4.2	His Own Story	30
		4.2.1 Early Experiences	30
		4.2.2 Talent	31

ix

	4.2.3	Superconductivity, Bardeen	31
	4.2.4	Cooper Pairs	32
	4.2.5	Schrieffer	33
	4.2.6	The Paper	34
	4.2.7	Approximations	35
	4.2.8	Breaking the News	35
	4.2.9	Contributions from Other Scientists	36
	4.2.10	Present Problems (2001)	37
	4.2.11	In Between?	38

5 John Robert Schrieffer: The Microscopic Theory of Superconductivity 41
- 5.1 Biographical Notes 41
- 5.2 His Own Story 43
  - 5.2.1 Early Inspiration 43
  - 5.2.2 At MIT 44
  - 5.2.3 Ambitions 44
  - 5.2.4 First Publication 45
  - 5.2.5 Thinking Big 45
  - 5.2.6 Success! 46
  - 5.2.7 Ultrasound 47
  - 5.2.8 Transitions 48
  - 5.2.9 Finished Work 49
  - 5.2.10 Josephson 49
  - 5.2.11 Rating Superconductivity 50

6 Ivar Giaever: Single Particle Tunnelling: Confirming the BCS-Theory 53
- 6.1 Biographical Notes 53
- 6.2 His Own Story 55
  - 6.2.1 Background 55
  - 6.2.2 Applying for the University 55
  - 6.2.3 Interests 56
  - 6.2.4 Lodging and Job 57
  - 6.2.5 Quantum Mechanics 59
  - 6.2.6 Tunnelling 60
  - 6.2.7 Reaction on the Paper 61
  - 6.2.8 The Josephson Effect: Seeing, but not Recognizing 63
  - 6.2.9 After Superconductivity: Biophysics, and Some More... 63
  - 6.2.10 Vision? 64
- Reference 65

7 Brian D. Josephson: Cooper Pair Tunnelling: The Josephson Effects 67
- 7.1 An Abbreviated Account Based on an Interview and Available Literature 67

Contents

8 Philip W. Anderson: Superconductivity from a Broader Perspective . 73
- 8.1 Biographical Notes 73
- 8.2 His Own Story 74
  - 8.2.1 Early Influences 74
  - 8.2.2 Career Choice, Family and Politics 75
  - 8.2.3 Electronics Physics and Harvard 76
  - 8.2.4 Encountering Superconductivity 77
  - 8.2.5 Cooper 77
  - 8.2.6 Order Parameter 78
  - 8.2.7 1959 79
  - 8.2.8 More He3, and Phonons 80
  - 8.2.9 Concepts in Solids 81
  - 8.2.10 The Josephson Effect 81
  - 8.2.11 Kondo Effect and the Renormalization Group 83
  - 8.2.12 Resonance Valence Bond (RVB) Theory 84
  - 8.2.13 The Situation as Seen in 2001 85
  - 8.2.14 The Future, as Seen in 2001 86

9 Pierre-Gilles de Gennes: The Orsay Group on Superconductivity . . 89
- 9.1 Biographical Notes 89
- 9.2 His Own Story 90
  - 9.2.1 Early Days 90
  - 9.2.2 Education 91
  - 9.2.3 Teachers and Masters 92
  - 9.2.4 PhD 93
  - 9.2.5 Becoming a Theorist 93
  - 9.2.6 Experimental Approach 94
  - 9.2.7 BCS and the Orsay Group 95
  - 9.2.8 Liquid Crystals 96
  - 9.2.9 Main Achievements at Orsay 97
  - 9.2.10 Why Leave Superconductivity? 97
  - 9.2.11 High-T_c 99
  - 9.2.12 Popular Lecturing 100

10 Johannes Georg Bednorz: Discovery of Cuprate Superconductors . . 103
- 10.1 Biographical Notes 103
- 10.2 His Own Story 105
  - 10.2.1 Path to a Scientific Career 105
  - 10.2.2 A Student's Experience of the "Real Word" of Research . 105
  - 10.2.3 PhD Studies at the ETH and First Encounter with
           Superconductivity 106
  - 10.2.4 New Project 107
  - 10.2.5 Risk 107
  - 10.2.6 Cu-Components 108
  - 10.2.7 Discovery! 109
  - 10.2.8 Identification 110

		10.2.9 The Meissner Effect Paper 111

           10.2.9   The Meissner Effect Paper 111
           10.2.10 Reactions 112
           10.2.11 Visiting the German Physical Society 112
           10.2.12 Nobel Prize 114
           10.2.13 Applications 114

11 K. Alexander Müller: Discovery of Cuprate Superconductors 117
    11.1 Biographical Notes 117
    11.2 His Own Story 119
        11.2.1 Background 119
        11.2.2 At ETH 119
        11.2.3 $SrTiO_3$ and Superconductivity Preliminaries 120
        11.2.4 Yorktown Heights: Learning by Doing. Conventional
               Superconductivity 121
        11.2.5 The New Beginning: Jahn-Teller Polarons 122
        11.2.6 The Discovery 123
        11.2.7 Definite Proof, Growing Attention 124
        11.2.8 Recognition and Priorities 124
        11.2.9 The Future, as Seen in 2001 125
        11.2.10 Concluding 126

**12 The Anderson-Higgs Mechanism for the Meissner Effect in
Superconductors** 129
    A. Sudbø

13 Concluding Remarks 133

Index 137

Photo Credits

Figure 1.1: Kristian Fossheim
Figure 2.1: nytimes.com *Permission requested*
Figure 3.1: Kristian Fossheim
Figure 4.1: Kristian Fossheim
Figure 5.1: National High Magnetic Field Laboratory (NHMFL)
Figure 6.1: Kristian Fossheim
Figure 7.1: Kristian Fossheim
Figure 8.1: Kathleen Blumenfeld
Figure 9.1: Françoise Brochard-Wyart
Figure 10.1: IBM Zurich Research Lab
Figure 11.1: Kristian Fossheim

Chapter 1
Introduction

Fig. 1.1 Bust of Heike Kamerlingh Onnes in Leiden

Physics is a science which aims at answering the big mysteries in Nature. Physicists have always been attracted by the greatest challenges. But sometimes even the most demanding problems reveal themselves little by little. On the 8th of April 1911 a discovery was made through an apparently simple experiment in a glass flask of very special design in a physics laboratory in Leiden, Holland. The experiment set in motion a series of events with few parallels in the history of science. But physics was far from ready for the advent of superconductivity, the enigmatic phenomenon which Heike Kamerlingh Onnes and his student Gilles Holst had just observed. Today, more than a hundred years later, after great scientific research efforts and big investments, and after many impressive scientific and technical breakthroughs,

K. Fossheim, *Superconductivity: Discoveries and Discoverers*,
DOI 10.1007/978-3-642-36059-6_1, © Springer-Verlag Berlin Heidelberg 2013

a cloud of mystery still hovers over aspects of superconductivity. Nature continues to play her elusive game with the best minds of physics.

It seems right, after passing the 100 year milestone, to take stock of the intellectual property upon which we stand in this field, and from which basis scientists launch further expeditions into the remaining enigma. Physicists' fascination with superconductivity prevails, and continues to attract new generations.

The year 1911 would turn out to be a great year in science history for an additional reason: The discovery of the atomic nucleus by Rutherford. Subsequently, the first model of atomic structure, the Bohr model, followed in 1913. The impact on science would be tremendous. 1911 will forever be a year hard to match in the annals of science.

It is not unusual in science history that an apparently simple observation opens a Pandora's box with wide-ranging consequences. H. K. Onnes studied electrical resistance in a metallic wire, hardly something that could change the world, you should think. In 1911 it was already known that electrical resistance in metals diminishes gradually and continuously as temperature is lowered more and more below the ambient. This fact had been carefully established by recent research, not only in Leiden. But Leiden had established itself as one of the central research arenas in the new field of low temperature physics, in a combination of curiosity driven search into new territory and development of fabulously sophisticated glass blown cooling devices. When H. K. Onnes and his team, in 1908, after many years of systematic efforts managed to condense the noble gas helium, the path was laid for unprecedented study of the low temperature properties of matter; gases, liquids and solids.

A problem which had been much debated at the time was what would be the ultimate low temperature behaviour of electrical resistivity on approaching zero degrees on the Kelvin scale. How low could the resistivity ultimately become? Would resistivity continue to decrease, and gradually vanish for all practical purposes? Or, would the current carriers eventually "freeze," or "stick to the atom" like some thought, thus preventing the charge carriers from participating in electrical conduction, forcing resistivity to increase again?

What H. K. Onnes and coworkers discovered, was something entirely different from both of these alternatives, and completely surprising: Resistance- and hence resistivity- in solid frozen mercury metal filaments vanished abruptly at about 4.2 K degrees above absolute zero, or at about minus 269 degrees centigrade, and remained zero at all lower temperatures. This phenomenon was called superconductivity. The temperature where it happens defines a dividing temperature which, as it would later turn out, is characteristic of each metal, and is called the superconducting transition temperature, T_c. In the pure metals of the periodic table, T_c would typically be below 10 K. As years went by, most but not all metals were found to be superconducting at low enough temperatures. Famous examples of non-superconductors, paradoxically it seemed, were the best metals like gold, silver and copper. Soon also a great variety of metallic alloys were found to possess the superconducting property. But no explanation could be found at the time.

1   Introduction

It would be wrong to say that the world of science stood in awe of the new discovery. When H. K. Onnes received the Nobel Prize in physics in 1913, superconductivity was not even mentioned.[1] Rather, the emphasis was on Onnes' great feat in low temperature science and technology leading to condensation of the highly volatile inert gas of helium. It would later turn out that yet another important property of superconductors had still to be discovered. Nature reveals its secrets only when the appropriate questions are asked through precisely designed experiments. As is often the case, the problem was to know which question to ask.

It would take another 22 years before that next step was achieved, in 1933, when the deeper nature of superconductivity was revealed in a magnetic experiment by Walther Meissner and Robert Ochsenfeld in Germany. Before discussing that experiment, let us first sidestep a little and recall some simple facts: The most common metals, like lead, tin and aluminium, are classified as very weak paramagnets. This important characteristic is due to the fact that although electrons have the ability to align their magnetic moments with an applied magnetic field, and thus reinforce an externally applied field, only a very tiny fraction, those with the highest kinetic energy, are allowed to do so in a metal. This is due to the lack of available quantum states for most electrons into which they can accomodate if their magnetic moment is turned parallel to the field. Therefore, the number of electrons oriented parallel and antiparallel to the field, respectively, are almost equal, and the magnetism of the "gas" of freely moving electrons in a metal is almost zero. This is what is characterized as weak paramagnetism.

The second aspect of superconductors, discovered in 1933, came just as unexpectedly as the sudden loss of electrical resistivity in 1911. A piece of metal was first held in the normal state above T_c, while its entire body was permeated by an externally applied magnetic field from the solenoid in which it was located. The resulting magnetic field inside the sample was then very nearly the same as that outside, as described above. The sample was then cooled through the critical temperature T_c. On passing T_c, it was recorded that the magnetic field inside was suddenly and completely expelled. Hence, by lowering the temperature by just a small fraction of a degree, the material changed its magnetic character completely, from weakly paramagnetic above T_c, to a state of complete screening, with no magnetic field in the body below T_c, i.e. perfect diamagnetism. This must have been caused by the sudden creation of an opposing field which exactly cancelled the applied field inside. This remarkable behaviour, never observed before, is referred to as the *Meissner effect*, a phenomenon which ranks among the greatest theoretical challenges ever encountered in the history of physics. It was demonstrated that this constituted a new thermodynamic state, and that it was *not* a consequence of infinite conductivity. The deeper nature of the Meissner effect as a realisation of the Higgs mechanism was discovered almost 30 years later by Anderson, as told by him in the Anderson chapter of this book. Further comments on the Higgs mechanism are given in Chap. 12.

---

[1] The citation for the Nobel Prize to H. K. Onnes in 1913 does not mention superconductivity explicitly. But there may be reason to argue that this was due to the short time between the discovery and the deadline for nominations. This was pointed out to the author by Tord Claeson.

Henceforth, zero resistivity and the Meissner effect were referred to as the two distinguishing characteristics of the superconducting state of metals. Only if both of these could be observed, would a material be counted as superconducting. It was further realized that since the Meissner effect was a persistent phenomenon caused by spontaneously created screening currents near the surface, the presence of the Meissner effect implied zero resistivity. The Meissner effect, or perfect diamagnetism, therefore is the true defining property of superconductivity. Only when this effect is observed, can one claim to have observed superconductivity.

The Meissner effect is named after professor Walther Meissner, who lead the experiment. It should rightfully be called the *Meissner-Ochsenfeld effect*, including the name of the student, but practice has been mostly to use the shorter name.

An important step was made by Fritz and Heinz London in 1935, when they proposed a description of the magnetic state of the superconductor in which the conduction electrons were divided into a normal part and a superconducting part. The screening of the interior of the sample against an applied magnetic field was described by the London equations as being upheld by a spontaneously created current of the superconducting electrons in a very thin surface screening layer, called the London penetration depth λ, its limiting value being less than a micrometer in simple metals. Such a phenomenological description was quite useful. Even so, the origin of the whole effect remained a mystery.

Onnes had quickly realized the potential for superconductors to replace conventional electromagnets since their current carrying capacity seemed enormous. To his disappointment, only quite weak magnetic fields could be created by superconducting solenoids he had available. Many properties of superconductors were not yet known, and lots of superconducting metals had not been discovered. It would take about 50 more years before useful, strong electromagnets could be made. The underlying limitations were due to limiting values of critical current density and critical magnetic field, the upper limits to how large currents and magnetic fields superconductors could tolerate. This called for the study of phase diagrams where such quantities were measured vs temperature. The field of superconductivity was growing ever wider.

Physics is a science which moves forward in an intimate interplay between experiment and theory, each advancing the other, in alternating steps. After many years of experimental progress, theoretical insight was lagging behind experiments. The greatest theorists in the field at this time were found in the Landau group in Moscow. Landau had already been engaged in work on the penetration of magnetic field in superconductors. He had also formulated a general phenomenological theory for systems which undergo continuous phase transitions between thermodynamic states. Applications of this theory require identification of a special parameter, the order parameter, which is different from case to case, and contains the essence of the problem. For the mathematical procedure to work, the order parameter must be small, or vanishing at T_c, and grow gradually on lowering the temperature. In the case of superconductors, the fraction of superconducting electrons could be seen as such a parameter. This was the situation when Vitaly Ginzburg and Lev Landau applied the Landau theory to superconductivity in 1950, and thus gave science a tool which has been of enormous importance in all the ensuing years. When Ginzburg received the Nobel Prize 53 years later, in 2003, it represented the ultimate recognition of the importance of his work with Landau.

The Ginzburg-Landau theory very soon found important applications. Alexei Abrikosov, also from the Landau school, had become much interested in the magnetic properties of superconductors, mostly through experiments carried out in thin films by one of his colleagues. He felt that the Landau theory was a remarkable tool. The results he obtained already in 1953, were unexpected. In fact, they were so surprising that his revered and respected boss, Landau, did not believe they were correct, and hence would not allow their publication. What Abrikosov had found, was a periodically ordered magnetic field penetration in superconductors, narrow lines of quantized magnetic flux, what has later been named the Abrikosov lattice. This was very different from magnetic structures studied by Landau before in the simple metals mentioned above, and Landau's refusal came because Abrikosov could not give a simple physical argument for his finding.

However, after Feynman's work on superfluid liquid helium, a parallel to superconductivity, where a similar effect was found, Landau gave in. Abrikosov's work introduced a new kind of superconductor, a "superconductor of the second kind," nowadays mostly called "Type II superconductor," as distinct from "Type I," which Onnes and many others, including Landau, had worked on before. Abrikosov's work became extremely important for future applications of superconductors. As an example, modern MRI would not exist without the knowledge and technology Abrikosov's work contributed to. Other examples are levitation technology for trains, high power energy transfer lines, and the Large Hadron Collider at CERN, Geneva, where the Higgs boson for particle mass was recently discovered. The Nobel Prize in 2003, shared with Ginzburg, came late, but was much deserved.

Even so, a phenomenological theory does not explain the underlying mechanisms of superconductivity. This was what Bardeen wanted to do when he carefully set up his team with two young and bright physicists at University of Illinois at Urbana, what later became known as the BCS team of John Bardeen, Leon Cooper, and John Schrieffer. These three turned out to be a kind of star team seldom seen, and their work gave a huge breakthrough in superconductivity research. Cooper first found a mechanism for electron pairing in 1956, known today as Cooper-pairs. Since the effect of pairing was to lower the energy of freely moving electrons in a lattice of metal ions, this was a very promising step. With Schrieffer's additional idea to write down the quantum state of the whole system of electrons, many properties of superconductors could be predicted. After much hard work, the BCS theory was published in 1957. A central piece of the theory was the prediction of the size of an energy gap due to Cooper-pairing. This gap could subsequently be determined by several experimental techniques, and agreed with predictions. Also, the density of states was predicted, and later confirmed. BCS gave a very complete and realistic description of the phenomenon of superconductivity. The essence was the coherent quantum state made possible by the pairing mechanism, which in superconductors was due to the exchange of lattice vibrations, phonons. Schrieffer often emphasises that the BCS theory has a much wider span of applicability, including atomic nuclei and neutron stars, a strong theory, with predictive power.

In 1960, Ivar Giaever, an employee at the research laboratory of General Electric in Schenectady, New York was taking physics courses at Rensselaer Polytechnic Institute in Troy, where he heard lectures on superconductivity. Having done experiments on tunnelling of electrons through thin normal metal films, he realized he could modify the tunnelling characteristics if he performed the experiment on superconducting films. He so did, and thereby established thin film tunnelling in superconductors as a new and exciting tool in the investigations of superconductivity. The experiment gave a very direct answer to the question of the size of the energy gap due to Cooper-pairs in the BCS theory. Even better, it gave a precise graph of the BCS superconducting density of states. His work was seen by the Nobel committee as the ultimate confirmation of BCS theory. For this work he received the Nobel Prize in 1973, the year after the BCS team.

Giaever shared the Nobel Prize with the young English physicist Brian D. Josephson, who had carried the subject of electron tunnelling in superconductors one big step forward. In this case theory was again ahead of experiment. Under the influence of a series of talks by Phil Anderson, and inspiration from Brian Pippard, his thesis adviser, and with knowledge about the experiments of Giaever, the young student, only 21–22 years old, did his life's masterpiece when he predicted the tunnelling properties of very thin superconducting films. The Josephson effects comprise several physical effects, the most astonishing one being the transmission of a DC superconducting current at zero applied voltage. This effect is driven by a difference in phase of the superconducting wave function between the two sides of the film, which has an effect similar to an applied voltage. Furthermore, if in addition a voltage is applied, a microwave field is radiated. In practical terms, the main importance of the Josephson effects has been in making possible very sophisticated magnetic field detectors. The Superconducting QUantum Interference Device (SQUID) is by far the most sensitive detector of magnetic field ever made. It is widely used in the measurements of brain waves and other biomagnetic signals from the human body, as detector of radio waves, as voltage standard etc. The important difference between the Giaever experiments and the Josephson devices is that in the Giaever tunnelling experiments single particle tunnelling is responsible for the effects observed, while the Josephson effects owe their existence to Cooper pair tunnelling, implying that the superconducting wave function extends across the thin film barrier.

The BCS theory was in one sense an instant success. On the other hand there seemed to be problems with gauge invariance. While the BCS team was not worried about it, others were. One of them was Phil Anderson, at that time still at Bell Labs, who clarified the issue. He is known for his broad efforts in many areas of solid state physics, among them magnetism and superconductivity. He clearly inspired Josephson's work. With Kim he predicted how the magnetic vortex lattice discovered by Abrikosov could be "pinned," or immobilised, thus preventing energy loss during transportation of electrical current in a superconducting wire. These days, Anderson's claim to have discovered the Higgs mechanism during work on the Meissner effect in superconductivity is worth special attention.

Some of the best physicists form tightly collaborating groups or teams that have the character of a "school." We mentioned Landau above. Pierre-Gilles de Gennes' group at Saclay in Paris was such a school, referred to as the Saclay Group on Superconductivity. de Gennes is not famous for a particular discovery in superconductivity. Rather, his work was an inspiration for a generation of young physicists in superconductivity, and he was a discoverer in soft matter. Equally important, his unique lecturing style which brought him to meetings all over the world, and to high schools all over France, did a lot to promote science in general, and physics in particular in a wider context. Those who heard him lecturing, will never forget this great communicator of science.

Until 1986 a severe limitation of superconductivity was always present: A maximum of only about 23 K for the superconducting transition temperature T_c. All those who ever became interested in superconductivity have shared a common dream: That this fascinating phenomenon could one day be observed and used at room temperature. It would be one of the most wonderful gifts of science to the world. Can it happen? It surely will not happen without scientists who are ambitious and courageous enough to try to reach for the impossible. Two men who did, were K. Alex Müller and Johannes George Bednorz at the IBM Research Laboratory in Rüschlikon near Zürich. Their work rounds off the story of the great scientists and discoveries in this book. Their discovery, against all odds, of a new class of superconductors with higher T_c, raised the hopes of thousands of scientists all over the world. "The Woodstock of physics," as New York Times named the first international meeting on the new subject in New York in January 1987, is a unique event in the history of science. At this moment, 25 years later, we look back at these events, and experience here how extremely exciting and promising science can be, and how demanding and challenging it is for those involved.

Chapter 2
Vitaly L. Ginzburg: The Ginzburg-Landau Theory of Superconductivity

> *"So, I introduced as the order parameter some macroscopic psi-function. And I came to Landau with this idea. He agreed, and we began. We immediately had this equation for the free energy with psi as the parameter, and afterwards we worked out all the rest; and that is the history."*

Fig. 2.1 Vitaly L. Ginzburg

2.1 Biographical Notes

Vitaly L. Ginzburg (1916–2009) shared the Nobel Prize in physics for 2003 with Alexei A. Abrikosov and Anthony J. Leggett *"for pioneering contributions to the theory of superconductors and superfluids."* His most famous and most important scientific work is the theory he developed with Lev Landau, the Ginzburg-Landau theory of superconductivity.

Vitaly Ginzburg set a firm stamp on several areas of physics in the Soviet Union. His research spanned a wide range such as: Classical and quantum electrodynamics, Cherenkov and transition radiation, propagation of electromagnetic waves in plasma, radio astronomy and syncrotron radiation, cosmic-ray and gamma-ray astrophysics, light scattering in crystals, theory of ferroelectrics, superfluidity and superconductivity. He was a co-inventor of the first Soviet H-bomb concept, with Sakharov. His Jewish family background, and charges against his wife, gave the family serious problems during the Stalin era, in spite of the fact that he received both the Order of Lenin and the Stalin Prize. "Stalin went totally insane," he wrote in his autobiography for the Nobel e-Museum. In his later years he voiced strong criticism against the Russian Orthodox Church. One news agency wrote upon his death: "Despite his age, Ginzburg remained active as a scientist and public figure. He also was a staunch believer in the global triumph of democracy and "secular humanism" to help overcome such threats as Islamic terrorism, poverty and AIDS."

In 1943 Ginzburg started work in superconductivity, trying to follow up Landau's work in superfluids which in its turn had been inspired by Kapitza's discovery of superfluidity in helium. First he worked on the thermoelectric effect. Eventually his interest focused on the application of Landau's general theory of second order phase transitions. His first application of this theory was in ferroelectrics, where he used polarization as the order parameter, and established the famous Ginzburg criterion for the validity of the Landau expansion. Superconductivity was a far less obvious case. He wanted to expand the energy in the superfluid density. But in quantum mechanics the density is the square of the wave function ψ (Greek letter *psi*). So he had to use the square of the still unknown ψ-function for the density. Hence the energy was expanded in a series in even powers of ψ. Out of modesty, Ginzburg preferred to call their theory "*psi*-theory" instead of Ginzburg-Landau theory. This theory has become monumentally important in superconductivity. It is usually applied as a mean field theory, but computationally it can be generalized to include fluctuations, and to also treat dynamical problems in superconductivity. Its wide applicability in high-T_c superconductivity has come as both a surprise and a blessing to this field where the coherence length is so short that initially there were serious doubts as to the validity of the Ginzburg-Landau theory in such cases. Theoretical progress in the field of high-temperature superconductivity, on the microscopic origins of the phenomenon, has been slow. It has been one of the major outstanding issues in physics for more than two decades, since the discovery of cuprate superconductors in 1986. However, the Ginzburg-Landau model has been enormously fruitful in uncovering and understanding the plethora of novel vortex phases that can appear in extreme type-II superconductors, such as the high-T_c cuprates, where disorder and thermal fluctuation effects are pronounced. This is extremely important for intelligent engineering of superconductors for large-scale applications.

It would be fair to say that the Nobel Prize to Ginzburg in 2003 was extraordinarily well deserved, and much overdue. Ginzburg has, in addition, received a number of other awards and honours.

2.2 His Own Story

2.2.1 Early Years

I was born in tsarist Russia in 1916. At that time in Russia we even used the old calendar, the Julian calendar. So according to the Julian calendar I was born on the 21st of September; on the 4th of October according to the new calendar. My father was an engineer, and my mother was a medical doctor. We lived in Moscow, and my

2.2 His Own Story

mother, very unfortunately, died in 1920 when I was only four years old. I was the only child in the family. My mother's younger sister then started to live with us and did everything she could for me. All my life I was living in Moscow, except for two years during the war, when we were evacuated to Kazan.

My life was far from privileged. Before the revolution my father had a flat with four rooms, and after the revolution two extra families were put in our home. Our flat was a so-called communal room. Also, I remember that around 1921 even children had to eat the meat of dogs. And we never eat dogs in Russia, so conditions were not good in general. In Moscow conditions were better than outside, but in general conditions in Russia were very bad, indeed. In Moscow it was better than in other places, and we didn't starve in any case. This was the situation.

2.2.2 Education

I had only four years of public school, according to a decision of my father, skipping the first three forms. And when I finished school in '31, higher education was abandoned. In Russia every few years there is a change in the system. So in '31 it was thought that everybody who had finished seven years—which I hadn't—had to go to a special school to be taught to be a worker. But I didn't go there. I chose to work as a technician in a laboratory. I worked in that laboratory for two years. I then very quickly educated myself in just a few months time, a very unfortunate situation which deprived me of normal learning development, as I have discussed in my Nobel autobiography. In spite of this, I finally entered the Physics Department at Moscow University in '33, and finished in '38.

But I have a very strong inferiority complex and decided that I was unable to be a theoretical physicist. I am not very good in mathematics, and theoretical physicists have to do a lot of mathematics. At the university, when I was registered for military service in '38, but without being sent to camp, I did optical work where I spent much time in a darkroom. I had a feeling that this wouldn't be very good for me. So I began to theorize, and I invented some possibility in quantum electrodynamics. I went to see professor Tamm. He was very friendly and told me that I had to read several books, papers etc. I was very quickly able to come up with several interesting results. I left optics, and from '38 I became a theoretician. I defended my PhD thesis in 1940 and moved to the Physical Institute of Academy of Science of the Soviet Union, now called the Academy of Science of Russia. And I have worked in this place already 63 years, except that in '41 we had to evacuate to Kazan for two years.

2.2.3 War Years, Tamm, and Sakharov

These were war years, but I stayed out of the war quite by chance, actually. It is quite interesting. Let me explain that. First of all, in Russia at that time there was a directive from the authorities not to take into the forces people who had no military education, and at the same time, had finished the university. I had no papers that I should not be drafted. I had always expected to be drafted, but the fact was that I wasn't, apparently due to medical classification given by the doctor who examined me. I certainly did not make any effort to avoid being sent into the war machine. Many of my classmates had to go, and many of them lost their lives.

So in '42 I defended my second dissertation. You see, in Russia there are two dissertations, one like a PhD, it is called "Candidate of Science," and the second is "Doctor of Science." So I finished the second one and defended it in '42. And after that work, in the same institute, I became the deputy of professor Tamm, who was head of the department.

On a historical note, let me mention the following: Much later, when Tamm died in '71, I became head of the department. I had to be the head of the department, because in Russia it is important that a member of the Academy would be the head, and we had at the time only two members of the Academy, Sakharov and I. Sakharov was not suitable, because he already was a dissident. So that would be no good. So I became head of the department in '71 and I was head until '88, when I was able to leave the job because of age. Before this, in Russia there was no age limit, and when a limit was introduced, I used it and left because I didn't like to be head of the department. And this permitted me to finish that kind of work. Beginning in '68 I had a part time job as a physics chair at Moscow Physical Technical Institute. It is a teaching institute, so it was a part time job.

2.2.4 Superconductivity, Landau

Many colleagues believe I got the Nobel Prize for what people call the Ginzburg-Landau theory. I disagree with this interpretation. Before my work in superconductors, I first heard a report on superfluids from Landau. Kapitza had attracted him, and he made his very important paper about superfluidity. This paper, and his report, were given in '40 or '41, just before the war in Russia. And when the war began, we had some other obligations. Only in '43 did I begin to work at low temperatures, trying somehow to follow Landau. Landau had solved the problem of superfluidity, and tried to do something in the direction of superconductivity. All this is already published in this book I mentioned. It is only published in Uspekhi Fizicheskikh Nauk, where I am Editor in Chief. Already for five years it has been available on the internet freely at www.ufn.ru. You can find it all in Russian and in English, from '97–'98, about the history of my results in this field.

My first work in superconductors was connected with the thermoelectric effect. From the beginning of '44 I was able to find, in some collaboration, that the Landau theory of superconductivity is not complete enough, that it is not applicable in strong magnetic fields. With strong fields I mean fields compared to the critical field. And so I began to think about the possibility to generalize the Landau theory. I worked in many directions. As a theoretical physicist I was able to do other jobs too, so it was a slow process. But in 1950 it somehow culminated in this work by Landau and myself. We were able to build what I call the *psi*-theory of superconductivity. If you like, you may call it the Ginzburg-Landau theory. I would not like to demonstrate my immodesty by simply, in Russian, using my own name; it is not comfortable. And also there are other reasons. So I call it the "*psi*-theory of superconductors," but the name is not important.

2.2.5 Superfluids and Low Dimensional Superconductors

I generalized this theory for the case of superfluids, and this was done, not together with Landau, but together with Pitaevskii. Landau took no interest in the development, and after 1950 I worked alone on superconductivity and superfluidity. But when Bardeen, Schrieffer and Cooper did their extraordinary work, I lost somehow interest in superconductivity, because now we understood the secret. Having no secret any more, I was not enthusiastic. But I returned to the field in 1964, after the work of Bill Little, with what I think was an important contribution. Little, of course, was a pioneer. I respect Little's work very much, but Little proposed a one-dimensional system. He didn't know at that time, and I didn't know, that in one dimension fluctuations are large, and I proposed a layer situation, two dimensional, or quasi two-dimensional, and sandwiches etc. And we worked in this field, with some group until the Bednorz-Müller discovery, and we published a book "High Temperature Superconductivity," which is the only book on the subject published before the work of Bednorz and Müller. So I think I have made some contributions on that. But I don't say that we made high temperature superconductors, because we cannot predict that you must have cuprates, or something like that. But I think we have contributed somehow.

So in short, I have also done something on superfluidity in neutron stars. I have a paper about this, and all the time I worked. So I don't agree that my work for which I received the Nobel Prize was only the *psi*-theory. I don't know the intentions of the Nobel Committee, but I myself think that I received the prize for long and persistent work in the field during many, many years, not only for the work in 1950. Officially, they gave it for pioneering work in superconductors and superfluids. My own opinion is that it is for my long efforts. Possibly they had in mind Leggett, I don't know.

Since the discovery of high-T_c superconductors it is an important question if the Ginzburg-Landau theory—if you permit med to use this name—is gaining importance in this area. I'm not quite sure. In some sense it is, but you see, you must take into account that the Ginzburg-Landau theory it is a theory of mean field, as Landau's theory of phase transitions. And just in the case of high-T_c superconductivity, the coherence length is rather small, so fluctuations are large. So there are some limitations. I do not follow all work going on now, so I cannot judge if generalizations which are being carried out are good.

2.2.6 How GL Theory Came to Be

I have been asked about the precise events that led to the mean field theory of superconductivity. As you will understand, this is a delicate question. Abrikosov and I were asked by the Citation Index about the history of our work many years ago. He answered, but I refused to answer. Why would I refuse? It looks like I would prove that I was just a co-author of the paper. You see, Landau was definitely as a physicist of the highest mark, higher than I, and he was my teacher. So somebody can suppose that I was a student or a post graduate for whom Landau has done everything and given me something to do.

I have decided that towards the end of my life I have the right to explain everything as it happened. I cannot prove it, because it was between us, with Landau and me, but I will tell what happened, and it happened the following way: In the beginning of '44, and this is published, I observed that Landau's theory is limited. What is the limitation? That it is not applicable in strong magnetic fields, and also Landau theory gives negative surface energy between normal and superconducting phases according to this theory. So you must introduce an extra surface energy, and this extra surface energy is large. This is strange, and step by step I tried to find how to handle this problem.

But the final step was that I tried to explain the large surface energy using quantum mechanics. You see, I used Landau theory of phase transitions, and in '45 I worked out a theory of ferroelectrics using Landau theory of phase transitions and introducing polarization as the order parameter. So I knew the Landau theory of second order phase transitions. Now the question was how to apply Landau theory of phase transitions in superconductors. It is a question first of all of choosing the order parameter. In the beginning I used the density as the order parameter in superfluids, and also in superconductors, meaning here the density of superfluid and the concentration of superconducting electrons in the metals.

But in the Landau theory, as you know, the order parameter squared enters, in the first term. And of course the density wouldn't come in the first power of rho simply, and not in rho squared, so you must use as the order parameter *the root of the density*. And what is the root of the density? It is just *psi*. So I introduced as the order parameter some macroscopic *psi*-function. And I came to Landau with this idea. He agreed and we began. We immediately had this equation for the free energy with psi as the parameter, and afterwards we worked out all the rest; and that is the history.

But it is very important, from the first glance even, that it looks like my role in the Ginzburg-Landau theory is even greater than, even much greater than Landau's. But it would be completely false, because it is a theory of which the basis is Landau's theory of second order transitions, so it says formally what is possible to do. In any case I think it is absolutely fair to say just Ginzburg-Landau, if you like to use names, because the basis is his. This is not an important formal situation, but that is the story. But in any case I claim that in no case I was a secondary author in this development. The idea of the order parameter was mine, and all the preliminary work was mine, and all this is proved by literature. So that is the story.

What I say also is that Landau was a great figure, and it is strange therefore, and I can't explain it, why he had no interest in for instance the theory of superfluids near the *lambda* point. Strange that he hasn't applied anything near the *lambda* point in liquid helium, or in superconductors. Why, I cannot say; he was at the time possibly more interested in fundamental problems of particle physics. In any case, all papers that followed were my own, and it is not the case that I neglected Landau. Simply, he had no great interest.

I would give you very interesting example which is in my paper, and of course in the Nobel lecture. It is the question of charge. As you will remember, in the Ginzburg-Landau theory—permit me to—there is some charge e^*, and from the very beginning, I thought that this e^* is an effective charge. Why would it have to be the charge e of the electron? Well, it is a quasi-microscopic theory. But Landau disagreed, and as a compromise in our work published in 1950 it is so formulated that there are no reasons to assume that e^* is not equal to the charge e of the electron. But I wasn't satisfied, and found how to solve the problem, partly.

You know, there is the parameter *kappa*, and in this *kappa* enters the critical magnetic field, which is observed, and the penetrations depth, which is observed, and also e^*. And *kappa* enters in the surface energy the field of supercooling and superheating. So, analysing the experimental effects I came to the conclusion that e^*, the charge in our theory, is $2e$ or $3e$, and I came to Landau and told him the result. And now he tells the objection. Possibly he had this in mind even before, but he hadn't told me in the beginning.

But this time he mentions a very important argument against. The argument was the following: If I introduce some effective charge e^*, this effective charge can depend on temperature, on density, on impurities etc. It means that it can depend even on coordinates. And in this case the translation invariance of the theory is lost. It is impossible in quantum mechanics, to suppose that the eigenvalues are functions of coordinates, because then the eigenvalues are lost. I tried to do something, but was unable. With Landau's permission, in my paper—I quote him of course—I say what is the situation, that according to experiments it is better to take e^* equal to two electrons. But Landau objected, and I didn't see how to go out of these difficulties.

What is the solution? It is trivial. It is really $2e$, but universal. So, both of us were correct. I tell this story because it is really interesting. Such now trivial ideas, that it is pairing, have really not come to my mind, or his.

In fact, even Cooper was not first. Cooper's work came in '56 as we know. But Ogg proposed pairing already in '46, and I quoted his paper. Afterwards Schafroth proposed pairing in '54. Cooper's paper is more important because Cooper not only proposed pairing, but he showed some real model which gives this pairing, which was very important for BCS. Later, Gorkov strictly showed that from BCS theory follows the Ginzburg-Landau theory, so-called. That is the history.

2.2.7 Remarks About the Prize

I was asked by the Nobel foundation to write a long autobiography. I did, I wrote it in Russian, and it was translated and sent to the Nobel foundation. Also, I have written two books. I have written several books in physics, but I have two books in neo-physics you might say, something more popular about physics. One of them is "Physics of a Lifetime," published by Springer in 2001. The other one is published in Russian, but is now being translated. As far as the Nobel Prize is concerned, I know that the first time I was nominated it was together with Gorkov and Abrikosov, probably in '75. After this I know I was proposed several times. But I want to stress that I never asked anybody to propose me. I think that would be completely wrong to do. I do not know who nominated me this time, but in two or three cases somebody told me: "I proposed you." So I know I have been nominated. And I would like to add that already long time ago I concluded that I wouldn't receive the Nobel Prize. I had several reasons for this, and I felt absolutely not harmed. I would be harmed for instance if somebody would receive this prize for my work. But if nobody receives the prize, it would not be important for me. And I remember very clearly that on the 7th of October, when the prize would be announced, I was absolutely sure that I wouldn't receive it. I came to work, and I was writing a letter to my granddaughter. But this time the telephone rang from Stockholm: "We are calling from Stockholm, you have received the Nobel Prize." At first I thought it might be somebody joking. But he said I would share it with Abrikosov and Leggett. After this I understood that a joker wouldn't invent such a complicated award. So I telephoned my wife and told her I got the prize, and did nothing further. But after half an hour it came by telephone and radio, and the hullabaloo began. That is the story.

Chapter 3
Alexei A. Abrikosov: The Magnetic Structure of Type II Superconductors

> *"I went and read Feynman's paper, and understood that it was exactly what I had proposed two years before. And I came and said to Landau, "Why do you accept this from Feynman, and you didn't accept it from me?""*

Fig. 3.1 Alexei A. Abrikosov

3.1 Biographical Notes

Alexei A. Abrikosov shared the Nobel Prize in physics for 2003 with Vitaly L. Ginzburg and Anthony J. Leggett *"for pioneering contributions to the theory of superconductors and superfluids."*

Abrikosov was born in Moscow in 1928. Already at the age of ten he was convinced he would become a scientist. He graduated from high school at the age of 15, in 1943. He had great talents in mathematics, but enrolled at this young age as a student at the Institute for Power Engineering. Still only 17 years old he was accepted as a student by the great Lev Landau, who understood what talents were at hand. Not yet 18 he passed Landau's famous test, "the theoretical minimum," and stayed close to him for many years to come. Eventually, he did his PhD with Landau, and was later a postdoc in his group.

During his long scientific life, Abrikosov has explored successfully many fields, but mainly the theory of solids: superconductors, metals, semimetals and semiconductors. He is famous for the theoretical discovery of what he called *superconduc-*

tors of the second kind, later mostly referred to as *type II superconductors*, and their magnetic properties, where magnetic field penetrates the superconductor by quantized amounts as vortices, in periodic arrays, called the *vortex lattice*. This magnetic structure in type II superconductors now bears his name as the *Abrikosov lattice*.

His main discovery was published in 1957, but the results had already been achieved in 1953, without Abrikosov being allowed by Landau to publish them. The reason was that his boss, Landau, did not initially believe the theory, and was not convinced until he learned that Richard Feynman had published an article where quantized vortices in superfluids were predicted to drive the so-called lambda transition in helium from the superfluid liquid phase, called helium II, to the normal liquid phase. The importance of Abrikosov's discovery of the quantized magnetic structure in superconductors can best be characterized as huge. In 1962 Abrikosov, together with all Russian physicists, and the world of science, suffered the loss of the creative mind of their great mentor, Landau, who was very seriously injured in a car accident, and later died, in 1968.

Abrikosov has had a distinguished scientific career. Around 1960 he worked with Khalatnikov and Gorkov on various aspects of superconductivity. With Gorkov he discovered gapless superconductivity. With Khalatnikov he did much work on superfluid He^3 employing the Fermi liquid theory of Landau. Together with Gorkov and Dzyaloshinskii, in 1961, he published a widely used textbook, *Quantum field theory methods in statistical physics*. Abrikosov has been a very active teacher during almost all his career. He has held several different professorships in Russia, later also in the US and the UK.

In 1991 Alexei Abrikosov moved to the US and joined the Materials Science Division as an Argonne Distinguished Scientist in the condensed matter theory group of the Materials Science Division, where he continued to be active. In Argonne he has worked on the theory of high-T_c superconductors, properties of colossal magnetoresistance in manganates and, together with experimentalists there, discovered the so called *quantum magnetoresistance* in silver chalcogenides. Abrikosov has been elected a member of the National Academy of Science in the USA, and of the Russian Academy of Sciences, Foreign Member of the Royal Society of London and the American Academy of Arts and Sciences. He has received numerous Russian and international awards, and the Honourable Citizenship of Saint Emilion, France. He was awarded an honorary doctorate from the University of Lausanne, Switzerland.

3.2 His Own Story

3.2.1 Childhood in Moscow

Both of my parents were medical doctors. And for reasons that I don't understand, my mother told me that under no conditions must I become a medical doctor. I don't know why she said that, but nevertheless she did. So therefore I excluded a medical career from the very start. You know kids listen to their mother, without any doubt.

I think that in religion it is the same, the kids get the religion from their parents. That is the usual thing. But on the other hand I had no doubt that I would become a scientist eventually, so I read various books about great scientists and inventors. This was when I was maybe less than ten years old. I read books about Michael Faraday, Bessemer, Edison, about various kinds of inventors, and other people.

I had a kind of a dream, not a realistic one; it was if I were a hero of ancient mythology, then I would fight various kinds of monsters and so on. And I had two other such unrealistic dreams. One was to become a member of the Royal Society, and the other was to get the Nobel Prize. Again this was approximately the age of ten, and of course I didn't consider that to be realistic. With my consciousness I understood that it was absolutely impossible. Afterwards, both of these were fulfilled. But that I couldn't expect. It says something about the power of children's dreams.

On a similar note, there is the story of the Frenchman Jean-François Champollion who was able to interpret the Egyptian hieroglyphs. He studied Egyptian writing from childhood on, among them the Rosetta stone and its translations. The Rosetta stone is in the British Museum and it has scripts in Egyptian hieroglyphs, in Greek and in demotic Egyptian language. Therefore, although hieroglyphic script is very different from ancient scripts like Latin and so on, he managed somehow to guess it. He had such a dream, and he achieved it.

I grew up in Moscow, and I worked in Moscow most of my life. You may say I was privileged due to my parents, my father mostly. He got the Golden Star of Socialist Labour Hero, and the Stalin Prize. He was the vice president of the Academy of Medical Sciences, and a very well—known person. He was one of the rare scientists who after the revolution did not emigrate from Russia, and also he did not oppose the Soviet power. He was a neutralistic person. He said, "If we have some power which allows us to work, why should we interfere with its actions? We should be grateful that we have the opportunity to work." And that was his view always.

Because of all this he was called when Lenin, the founder of the Soviet state, died. He was called on and performed the autopsy of Lenin, and his assistant made the first sort of balsamation of the body in order that people who came from far away in huge numbers could actually see the dead man and say goodbye to the person whom they considered like a god. Many rumours exist regarding Lenin's body, and his brain, but I heard nothing from my father.

3.2.2 School and Education

I liked going to school, of course, and I had many friends. Also, I was very successful with my lessons, in particular when it came to mathematics. So mathematics was very easy for me. I would say I had an intention of becoming a mathematician, but I did not.

When the war started in Russia, I was 13 years old, so I was much too young to be drafted, and when the war ended in '45, I was still only 17 years old, and not yet of the draft age. But on the other hand I was very much advanced in my studies.

I graduated from high school at the age of 15, and then I went to the Institute of Power Engineering (IPE). From that place they did not draft students, they gave them a delay until they graduated. I went to that school during the war and then I transferred to Moscow University. They created special groups on nuclear physics and engineering. This was in 1945, when the Americans exploded their nuclear bombs.

3.2.3 Entering the Landau Group

I always had such an idea that while studying at the university I must simultaneously start work at some research institute. In the Soviet era university teaching was separated from research institutions. And so research was done in a different place. There was some research at universities, but much less than in the research institutes.

So already when I was at the Institute for Power Engineering I wanted to do some research. It happened that my mother knew a physicist by the name of Vul at the Lebedev institute in Moscow. He invited me to work in his lab, and so a few days a week I came to his lab and was working there on the physics of semiconductors and insulators. We were measuring the dielectric constant of barium titanate. And simultaneously I went to the university and to the IPE. My department was called the Electrophysical Department. My real goal was to study physics.

I chose that institute both because at first it delayed the draft to the army, and also because at my young age the university didn't want me. I was only 15 years old, and IPE took me. But when I went to the university and compared what we were learning and what they were learning, I found that they learned much more. I therefore actually took the notes from the lectures and the textbooks and prepared for some exams. And we arranged so that I could make it, and I could pass the exams for the university, for the Physics Department. I had what I would call a triple load, at the Power Engineering Institute I took courses, while carrying out experimental research with Vul at the Lebedev institute, and then at the same time I was also preparing for, and passing the exams at the Physics Department at Moscow State University in 1945.

But then, next year it was declared that the country needed nuclear specialists, and they created special groups at the Physics Department at the university, and these students would also have a delay of the draft, and they would learn nuclear physics in what I would say was some advanced manner. They first wanted us to graduate earlier but this didn't work out. But anyhow I didn't lose any time. I came to the university and had these ideas that I must combine it with some research work, and the only name I knew in physics—since I came from a doctor's family—was Kapitza. So I asked my father whether he knew Kapitza, because my father was a member of the Academy. And, yes, he knew Kapitza and spoke with him. Kapitza became interested when he learned that there existed such a young and able guy and he wanted to see me.

When I came to his institute he told me, "You are such a young person, and you must not become very narrow from the start; you must take a broader approach." And in this connection he spoke with two people, one was an experimentalist, the other was a theorist. The theorist was Landau. And so I spoke with both of them.

I was around 17 when I came the first time to Landau and rang at the door. His wife—she describes it in her memoirs—saw a boy, a small boy, and she said, "Who do you want to see?" and I said, "professor Landau." And she said, "You have to be mistaken." "No, no, I'm a student of the university and I want to see him." And we made the appointment. Landau gave me his program which he called the "theoretical minimum." It consisted of nine subjects, which I had to learn and to pass personally with him. And so I passed these exams.

Landau was actually, I understand now, very kind to me, but he had principles and the principles were hard, and therefore he should not show his kindness. That was always his way, and he would have the same requirements to everyone. But actually, since I passed successfully every exam, I was never turned down. So he was happy with me.

Then next, in order to pass the graduation of the university I would have to go to some nuclear plant or nuclear research institution. I wanted to go to Landau and become his PhD student, I should say, the Russian system is somewhat different from the West. So therefore, in order to achieve that I had to somehow get out of the nuclear group. And this was done, but it was very hard, very hard.

3.2.4 With Landau

But nevertheless, I managed. As I said, I successfully finished the theoretical minimum and became his post graduate student. Landau preferred to work alone. He never proposed a topic for research. But he needed the young people, they were welcomed very much. He needed them for several purposes. He needed them first of all because he hated reading other people's papers. And therefore all students periodically gave reports at his seminar about somebody else's work so that he could reproduce everything himself.

For a very long time I was the secretary of that seminar. I therefore brought some journals and he marked the papers which he considered interesting. I wrote cards and put them in some box. Then everybody could choose a card, and there was a group of people who were his pupils who participated in those seminars.

In alphabetical order they had to give the reports. When a person's time approached he searched in the box, chose a card and studied that particular work and reported it at the seminar. And although I was the secretary, nevertheless I had to do the same kind of thing. This was the first thing which he needed, but he also expected that we would somehow invent various topics for research and inform him on our progress. And so his knowledge would increase. But he never tried to put his name on what we did. It was very difficult to convince him to put his name on anything, I would say. Of course he would put it on the work he did himself, but then he had only his own name. I have something like four or five papers together with him.

It happened that I and Khalatnikov, my colleague, first got some ideas we were developing, and then Landau got interested, but he was not actually an expert in these methods. Therefore we taught him these methods for a month or more. And eventually he understood what we were doing, and he said that it was all rubbish, and then he thought about it. Then he invented some principles, and after that we were quite conscious and we applied these principles. And practically doing it, writing equations, solving them and so on, that was my task. What he had called rubbish was good for inspiration, but not good for really constructing a theory. And I agree with that. He had a good judgement.

3.2.5 Superconductivity: Discovery of Type II

Superconductivity I bumped into. Landau never invited me into that. As I said in my Nobel lecture, when I first asked him how to find a topic for my own research, his main advice was to talk to experimentalists. I did, and I had with me my university mate, who was in the same group as myself, Zavaritsky. He later died young with some type of cancer. We talked with the experimentalist and Zavaritsky did the main experimental work, checking the Ginzburg-Landau theory for the critical magnetic field of thin films. He did just that, and he got a beautiful agreement with Ginzburg-Landau theory and so it was quite well. However, his boss, Shalnikov, who was a perfectionist, said: "Your samples are just nothing, they are dirty. And you prepare them so that you take a drop of metal and you heat it up, and then the drop is evaporated and the fumes fly and condense on a glass substrate." "However," he said, "your glass substrate is kept at room temperature, pretty warm, and when the atoms come onto the glass, they can move. There they agglomerate, and what you have is actually a plane covered with drops, instead of a continuous film. In order to prevent that, you have to keep your glass substrate at helium temperature and never heat it before you make your measurements."

Now, that was a little bit hard, but still Zavaritsky managed to do it, and when he did, there was absolutely no agreement with Ginzburg-Landau theory! That's how it was. And so he said, "Look, now I cannot get agreement with Ginzburg-Landau." Then we started to think, both of us, to think what can be the reason. The Ginzburg-Landau theory looked so beautiful, it was much better than anything existing at that time. So it couldn't be wrong. That was how convinced we were. So therefore we had to search for opportunities within that theory, and that was what we did. And we found a possibility there.

And since it was theory, I did a corresponding comparison, and when I calculated and compared, it fitted completely. What had to be done was the following: You express the quantities entered in the Ginzburg-Landau equations in dimensionless way, introducing corresponding dimensionless units, for instance, instead of magnetic field, introduce magnetic field divided by some value. If you do such a procedure, then all material units disappear from the equations. They become universal. Except for one quantity. There remains one quantity which was later called the Ginzburg-Landau parameter, with the Greek letter *kappa*. The value of this parameter can be defined from the intermediate state, a periodic structure of distribution of magnetic and non-magnetic regions in the superconductor depending on the sample geometry.

This so-called intermediate state was known long before the Ginzburg-Landau theory. Ginzburg-Landau theory was published in 1950, while the Landau theory of the intermediate states had been published already in 1937. So that was known, but there existed some quantity which Landau introduced quite empirically: The surface energy. The Ginzburg-Landau theory predicted the surface energy. And the surface energy had a direct connection with the Ginzburg-Landau parameter *kappa*. So therefore knowing surface energy, which was defined from the period of the intermediate state, you could define the parameter kappa, and for conventional superconductors it always appeared to be very small.

And that simplified the theory considerably. The main idea of the Ginzburg-Landau theory is that they introduced as the order parameter some type of wave function in the Landau theory of second order phase transitions. That was the central point of that theory. This was a big change of thinking.

The application of the wave function started just with my work. The phase played no role in the work of Ginzburg and Landau. Of course, in principle they kept it in. In principle the wave function is a complex quantity, but in all the calculation, when they calculated the surface energy, the critical field, the critical current, there was no phase. I started with, first of all, solving the problem of critical field of thin films with large values of the GL parameter *kappa*. They had considered only small values. When I considered large values I got agreement with experiments.

The possibility that the order parameter would be suppressed by the magnetic field, so that it wouldn't be constant when you turn on the field, was not considered at this stage because in thin films it was constant. It changed with the film but it was constant over the whole thin film. So at this moment it did not require any phase, only the absolute value. Nevertheless, in this case it was written in the Ginzburg-Landau work that if the GL parameter *kappa* is large, the surface energy between normal and superconducting phase becomes negative. And that was an argument in their paper, not to consider large *kappa* values because negative surface energy would be unphysical, since everybody knew that there existed an intermediate phase in superconductors known at the time. This did not require considerations regarding the phase of the wave function.

However, it showed us that there really exists substances with big *kappa* and negative surface energy. *And that meant that there exists a new type of superconductor.* So Zavaritsky and I called them superconductors of the second group in our papers which were published in the same journal. For some reason, unknown to me, he did not want to write a paper together with me, and so we wrote two different papers. And they were published in the same issue of this journal, which was never translated into English and is totally unknown, I think. And that was the first time, I think, that the idea that there exists two types of superconductors was published.

Now, de Gennes attributes the discovery, the practical discovery of type II superconductivity to Shubnikov. The reason is that de Gennes did not know my paper, because my paper was never translated to English. I described this too in my Nobel lecture.

It was so that there was de Haas, and he was a very well known physicist. de Haas with the wife of Casimir, whose name was Casimir Jonker, published a paper in 1935 were they measured the magnetic properties of superconducting alloys. I don't remember what alloy it was, but they got a gradual transition from a superconducting to a normal phase, with gradual entrance of the magnetic field into the superconductor. That was the first paper on this subject. And they said that there exists a special state with two critical fields. That's what they got from their experiments. Now Shubnikov was a former student of de Haas. He looked tentatively at what de Haas was doing and decided to repeat this experiment, but with much better samples. de Haas had written that he attributed the behaviour I mentioned to inhomogeneities, that the superconductor consists of pieces with different critical parameters, so first some of them turn normal, and then the others.

This was in 1935. Shubnikov decided to repeat the same experiment, but with better samples. He wanted to get rid of inhomogeneities, and study homogeneous material. And so what he did was that he heated the samples for a long time, very close to the melting temperature. And after that he made an X-ray, and if the material was inhomogeneous, where different parts had different properties, he could expect that there would be lines in the X-ray spectra, which belonged to different pieces. He didn't find any of that, but his measurements were done at room temperature, for some reason, I don't know for what reason, but that is experimental stuff. He could not make X-ray photos at low temperatures.

So he got extremely smeared out curves, despite this homogenization. And so he wrote that he could not explain that in any way other than by inhomogeneity. Therefore he wrote, "We don't see inhomogeneity with the X-rays, however it nevertheless must happen, and probably it is due to precipitation of another phase at lower temperature." That was written in Shubnikov's paper, so therefore Shubnikov was not the first person who observed such a gradual transition for a superconductor to another phase. This was probably not known to de Gennes. Shubnikov definitely considered inhomogeneity, and therefore not another type of superconductor.

However, Shubnikov was accused of attempting to organize an anti-Soviet strike, and was arrested and immediately killed by the KGB. So he is a tragic figure. And furthermore I can tell you, there was this guy, Kurt Mendelssohn, quite well known. And Kurt Mendelssohn invented a description of this "inhomogeneity" which was called "Mendelssohn's sponge." It was described as a kind of sponge of a superconductor with higher critical temperature which was embedded in a superconductor with lower critical temperature.

He worked in England. He was a German who had emigrated to England. Now, after my work was accepted he so hated it, because my work undermined the Mendelssohn sponge, and therefore he did a very sly thing. Namely, it was known already that there is a different phase and that it appears in type II superconductors. And now he proposed to call this new phase with the vortices the "Shubnikov state." First of all Shubnikov had nothing to do with it, and secondly it was not *for* Shubnikov, it was *against* me.

So then I decided to see what happens in bulk type II superconductor. I started with the vicinity of the second critical field, because of what I already knew. So the question is: How does the transition happen? From the films I saw that it is a second order transition.

So therefore one can reason as follows: If you have a nucleus of some different phase, and the field is too large, then the nucleus will gradually disappear, and will decay. But if the field is too weak it starts to grow, and it will grow and increase with time. And therefore; a stationary nucleus that does not grow and does not decay must define the field which is the transition point. And such a field was actually described in the work by Ginzburg and Landau. Just for mere curiosity they found that, and that was considered an additional proof that large values of the GL parameter *kappa* are impossible.

However, I took a nucleus which could exist at higher fields, only one nucleus at some certain point. Then I imagined the following: It can appear indefinitely, so therefore you can derive the linear combination of such nuclei. And because I saw through it homogeneity of space, in an infinite sample, it must be periodic.

However, one can't write such a periodic function as a mere sum, you have to introduce phase factors, because the particular form of one nucleus depends on the calibration of the vector potential. The vector potential enters the Ginzburg-Landau equation, and therefore to do that, the only way was to write a linear complex function. And therefore I wrote that complex function, and it described what happened close to the second critical field.

Now I wanted to know what happens at lower fields, so here I tried many constructions, and afterwards I saw all of these constructions in different papers. But I was not satisfied. Landau agreed with some of them, but I was not satisfied. And then I thought, "Why must I invent something? I must analyse. The wave function that I got close to H_{c2}, maybe will teach me something." And so I found that it has zeros. I did not introduce these zeros, I did not do that on purpose. It somehow appeared by itself. And just thinking about that, I understood that there is no other way to compensate the growth of the vector potential. But in reality the magnetic field, in average, doesn't grow, and the absolute value of the wave function does also not grow. So therefore the growth of the vector potential has to be compensated. And it can be compensated only by the phase. And so here the phase came in. It is also in my Nobel lecture how really that compensation happens. Afterwards, in the so called gauge theory, they found that they have also no way of compensating the growth of the vector potential, except for these singularities. Then I also understood that the lower the magnetic field, the larger is the distance between these singularities. And in the limit of a small magnetic field I can see only one of them, and this one of them was the point vortex. After that, after I made the field theory, the structure and so on, I could calculate the magnetization as function of the magnetic field. Then this was easy to measure, and I could compare with the best experimental work. And that was Shubnikov's work. And when I compared my theoretical results with Shubnikov's results, it fitted perfectly. Note that it doesn't depend on pinning. Pinning becomes important only if you send currents through the structure. Then the vortices start to move, but if you measure static magnetization, nothing moves. So therefore pinning is absolutely of no importance, not interesting here.

3.2.6 Difficulties with Landau

Landau was not happy, and I had already experiences like that with Landau. When I brought him something that was unusual, he required some simple argument for the theory. When I made my PhD thesis, for example, it was on thermal diffusion in plasmas. I found in partially ionized plasmas which has ions of different charge, that thermal diffusion could change its size and reach huge values. He said; I don't believe that. And so he said; I will only believe it if you give a simple argument. So then I disappeared for a week and I tried to work it out. And I succeeded. I brought it to him, and he immediately accepted it.

But here I could not give him such a simple argument. Therefore he disagreed. This was quite unfortunate for the work. Now, at that moment we had an interesting problem in quantum electrodynamics, the behaviour of Green's functions at high energies, and I put my theory on superconductors of the second kind in the drawer, and there it stuck for four years. I understood that my theory was correct, and thought I will be able to convince him one day. At this point it was not the most important issue.

And at this time the world became interested in the work by Feynman on vortices in rotating helium. Since Landau constructed the theory of superfluids, dealing with helium, he believed he must know everything about liquid helium. So he read Feynman's paper and he came and he said: "Of course Feynman is right, and what we published with Eugene Lifshitz is absolutely wrong." They had considered a system of concentric cylinders in rotating helium. And so I went and read Feynman's paper, and understood that it was exactly what I had proposed some years before. And I came and said to Landau, "why do you accept this from Feynman, when you didn't accept it from me?" Landau said I had done something different. Then I gave it to him and he understood that it was exactly the same. His attitude was, "You should have explained it to me."

This paper of mine was translated into English, but nevertheless, it did not attract attention. And people were accustomed to what Mendelssohn published, the inhomogeneous model. Then, however, they found alloys with very high critical magnetic fields and started to investigate their properties. And so, after that there existed an experimentalist who worked in Grenoble in France, Bruce Goodman, an Englishman. He published his theory, which was something like an intermediate state. And then probably somebody drew his attention to my paper, and then he published another paper, very strange, where he gave a more thorough explanation of his own theory, and my theory. Then he compared both with experiments, and came to the conclusion that my theory fits the experiments much better than his own, and just to demonstrate that, he published his paper.

3.2.7 Acceptance

You understand that such things never happened, they never happened before, and never happened afterwards. So after that people started to refer to both papers, and after a while Goodman's paper was absolutely forgotten. So that was how it developed. Nevertheless, experimentalists did not believe in vortices and vortex lattices. Then came experiments with neutron diffraction by the group of Cribier in France, but still experimentalists did not believe it. Neutron diffraction was not actually, at that time, something what everybody knew. So they didn't really believe that neutrons could find the diffraction.

Only after two Germans, Essmann and Träuble, did decoration experiments in 1967 where you could see the lattice, could people get accustomed to the fact that there exists some periodic structure. After this my theory was accepted. This happened ten years after my work was published, in '67, 14 years after I did it. So when people say that I got the Nobel prize too late, I reply that it was very hard to get recognition. And I can tell you what I'm doing now, which is the theory of high temperature superconductor. I can explain everything, and nobody recognizes that. And that was all my life so. Therefore, I behave very peacefully, that's why I'm still alive, have no heart attack, because I consider it as natural that people don't want to understand my theories. I mean, they are very simple to understand, but people don't want to understand.

So now, in the sense, finally and gradually there was confirmation, and everybody agreed that I was right. And then people started thinking about pinning and various other aspects. Eventually the work on these type II superconductors, in particular on vortex lattices and even more so for high T_c vortex liquids, all became so popular that now such lattices are everywhere.

I should add that Anderson and Kim did their work in the early sixties, and they believed in me, but they were theorists. They believed my work.

Now you see people speak about vortex matter as some special stuff, which in some sense permits us to understand new problems. The vortex state and vortex lattice is really something beautiful.

3.2.8 Side Issues, the Bomb, KGB, the Prize

I had no contact with Sakharov; I was too young. Sakharov was also not in nuclear physics from the start. He joined it when it came to the fusion bomb, mostly. And it was so that he actually was a student with Tamm, Igor Tamm, who got the Nobel Prize for the Cherenkov radiation together with Pavel Cherenkov and Ilya Frank. And Sakharov told Tamm that if he would be allowed to participate in this work, he had some ideas. Then Tamm proposed to include him in this work and it was successful. So Sakharov was very active.

I cannot tell exactly what was the reason why the Soviets came ahead of the Americans. It is a bit complicated. One of the things was that it was Ginzburg who actually had the idea, not the Americans who relied mostly on deuterium and tritium for the fusion bomb. There were various complications because tritium was unstable, and so Ginzburg proposed lithium and hydrogen. Both of these substances were common and if combined, they could actually decay into helium nuclei in a fusion process, producing very large energies. But also some major difficulties were absent with that method. This was in the late 1940s. They had the Landau group performing calculations. I was at the same institute now. Landau wanted to attract me towards this work, but he didn't get the permission from the KGB. It is true that Landau had been jailed in '38. But he was released after some time, because Kapitza behaved very smart and insisted on it. Now it was a different situation. Already they had made some calculations, and Landau wanted to attract me to these calculations. But KGB did not permit it. And only afterwards I learned, although not with complete definiteness, that this was because of the brother of my father, whom I had never seen and even did not know about his existence. The revolution had happened when he was an assistant ambassador in Japan. He never returned to Russia. He stayed in Japan until the end of World War II, and then he moved to the US and died shortly after. His name was Dimitri Abrikosov. But he wrote very interesting memoirs. I got these memoirs much later, and found it to be an extremely interesting book. It was originally just a manuscript which was found at Colombia University by a specialist on Russia of that period, by the name of Larsen. That person edited the manuscript and published it under the title "Revelations of a Russian diplomat." That is a very interesting book, but somehow I didn't know that person, my uncle. I didn't know about his existence. Nevertheless the KGB considered me not suitable to work on the bomb.

Anyhow, the Landau group made their calculations, and that helped a lot. So therefore, when they made practical tests of the fusion process, everything worked exactly as predicted in these calculations. This was half a year in advance of the US. The US was working on a static, on-the-ground device. But the Soviet bomb was already transportable, thrown from a plane, and everything worked perfectly.

You might think that the propaganda machine of Stalin would have influenced me. My answer is it did not, even though I became a member of the communist youth. But so was everybody. That was a condition in order to get into a university. But I never entered the communist party. I am reminded of a saying; that a person who at twenty years of age is not a member of the communist party has no heart, but if he is still a member at the age of fifty, he has no brain.

To answer your question, how long I have been waiting for the Nobel prize, I can tell you that somewhere in the early seventies Professor Roald Sagdeev who was the head of Institute for cosmic research in Moscow, told me that he got an invitation to nominate a candidate for the Nobel prize, and he wanted to nominate me. That probably was the first nomination. So one can say I have been waiting for thirty years.

Chapter 4
Leon N. Cooper: The Microscopic Theory of Superconductivity

"Don't try to solve a complicated problem if there is a simpler one you don't understand."

Fig. 4.1 Leon N. Cooper

4.1 Biographical Notes

Leon N. Cooper shared the Nobel Prize in Physics 1972 with John Bardeen and John Robert Schrieffer *"for their jointly developed theory of superconductivity, usually called the BCS-theory."* The famous BCS-theory stands out as one of the major achievements in all of science in the 20th century. Winning the race for a theory of superconductivity has been compared to the race towards the discovery of DNA by Watson and Crick in 1953. Superconductivity was referred to sometimes as "the holy grail." It was, according to Phil Anderson, "where all the big guys wanted to be." One who was there already was John Bardeen who came to share the Nobel Prize for research on semiconductors and the transistor effect in 1956. From 1955 he would organize the final, successful effort to understand superconductivity.

Leon N. Cooper was born in New York in 1930. He attended Columbia University where he received his A.B. in 1951, A.M. in 1953, and PhD in 1954. During 1954–55 he was a member of the Institute of Advanced Study at Princeton. He held a post doctoral position as a Research Associate with John Bardeen at Urbana, University of Illinois during 1955–57, and served as an Assistant Professor at Ohio State University 1957–58. Since 1958 he has been Professor at Brown University.

In his own account, his interest in superconductivity began with meeting John Bardeen at Princeton in 1955. Until then he had no previous knowledge of the field.

His background was in field theory, exactly what Bardeen was looking for. His first task upon arriving at Urbana at the age of 26, was to learn the basics of superconductivity. He became convinced, as was Bardeen, that the essence of the problem was an energy gap in the single particle spectrum as evidenced by the exponentially decreasing heat capacity towards $T = 0$ K. From a lecture by A. B. Pippard he learned that the basic facts of superconductvity appeared to be simple. In this respect it was an advantage that the isotope effect had been established, while all the exceptions found later were not yet known. Therefore, the ionic vibrations, "phonons," seemed to play an important role, as had been discussed by Bardeen and Fröhlich. But first he made the important proof, later referred to as the *Cooper problem* and *Cooper pairing*: The instability of the electron gas—the fermion system—against formation of electron pairs in the presence of the slightest attractive interaction between two electrons.

An intense collaboration with John Bardeen and his young student Robert Schrieffer started, with the aim to develop a full theory for superconductivity. In 1957 their famous BCS-paper was published. The pairing due to the previously envisaged hypothetical attraction between two electrons was identified as an electron-phonon interaction, events by which electrons with opposite momenta and spin in a thin shell near the Fermi surface formed a short-lived binding. The effect of this, happening all over the Fermi surface, was to create a new ground state, the superconducting state. The new theory had all the right properties, the energy gap, the Meissner effect, the penetration depth, the coherence length, the isotope effect, the prediction for ultrasonic attenuation, the so-called coherence factors showing up in the NMR relaxation rate etc.

Cooper was appointed Professor at Brown University in Providence, Rhode Island in 1958 and has remained there since. He has later changed field entirely, becoming the Director of Brown University's Center for Neural Science, founded in 1973, to study animal nervous systems. The center created an interdisciplinary environment with students and faculty interested in neural and cognitive sciences towards an understanding of memory and other brain functions. Among his scientific papers in neuroscience is one by Bienenstock, Cooper and Munro (BCM). Most people in neuroscience know him as a coauthor of the famous BCM-paper. Professor Cooper holds a number of honorary doctorates.

4.2 His Own Story

4.2.1 Early Experiences

I was born in 1930 in New York City. I guess I got into science because I liked it. As you might have guessed, I was good at it and got reinforcement. And I went to good schools, and enjoyed what I was doing. I don't think there was a family history in the science direction, except maybe far back. I went to the famous Bronx High

School of Science, where actually, I worked on projects in biology, more precisely bacteriology. It may perhaps explain things I did later in my career.

The project for which I became a finalist in the Westinghouse Science Talent search, was to grow a strain of bacteria normally susceptible to penicillin, and make it resistant to penicillin. Or more resistant. And the strain of bacteria was *Bacillus subtilis*. Basically I just grew the bacteria in varying concentrations of penicillin. And I took the one that grew in the highest concentration, and repeated the procedure. Finally, I got variants that grew in higher concentrations. The next project was to figure out why, but I never got to that.

This is really a problem nowadays that people are very concerned about. I think people have a much better idea about possible molecular mechanisms now. At that time, I don't think anybody really knew. The fact that I was able to do it, could be a little warning signal, considering the overwhelming use of penicillin nowadays. It seemed reasonable to believe that there would be variation in the wild type in terms of how it could handle penicillin. And if you kept taking the ones that handled penicillin best, then it seemed reasonable. Actually, it wasn't so easy to do it because the new strains reverted very quickly to the wild type, but you have to remember that I was just a student, and I had to wash my own test tubes...

4.2.2 Talent

I compared myself to the rest of the class, and found I did very well, and so, science became a fascination, probably when I was in junior high school, when I had a little laboratory where I would make experiments. My laboratory was in a closet. I used to mix chemicals, make magnets and do photography. Actually, when I was in college I loved classics, literature and philosophy. So science wasn't my only interest, but it was the direction I decided I wanted to go.

My father didn't think it was the brightest thing in the world to do, but it was better than my previous choice, which was to be a fighter pilot. And then when I was in college, I had to worry about whether I should go into biology or physics. I chose physics. There had been teachers who inspired me. I remember a teacher in junior high school, and there were others. In particular at the Bronx High School of Science there was a woman who maintained the laboratory where you could do experiments. She was just a terribly nice woman. Probably the reason that I did biology rather than physics, was that they had a biology laboratory, they didn't have a physics laboratory. So I just worked there every afternoon.

4.2.3 Superconductivity, Bardeen

Superconductivity was probably mentioned in some kind of course I had, but I don't remember paying too much attention. I know it was mentioned in a book I studied from, a book by Boorse, a book on thermodynamics. So, I might have seen it there, but I was really introduced to superconductivity when Bardeen came to Princeton, the Institute for Advanced Study. He was looking for a field theorist who was willing to work on the problem of superconductivity. He had written to C. N. Yang and T. D. Lee, and they apparently both suggested me. So Bardeen came and talked to me, and I said, "You know, I don't know anything about superconductivity." I hadn't even had a course in what we then called solid state physics. Bardeen said, "That doesn't matter, I'll teach you everything you need to know. Don't worry about it." At the moment, I felt that I wanted to do something a little bit different, so I decided to try it. At this stage I was already a post doc.

My thesis had been on nuclear physics, mu-mesonic atoms. mu-meson/lead atoms had an unexpectedly large energy in transitions from 2p from 1s levels. This led Ernie Henley and me to the conclusion that the charge distribution of the nucleus was smaller than had been thought previously. And that the charge radius was smaller than the neutron radius. I did my thesis with Robert Serber, who was Oppenheimer's lieutenant at Los Alamos. In fact, I teach in a course based on the play *Copenhagen*, that you may have seen, in which we refer to Bob Serber's famous introduction to nuclear physics—must reading for all the physicists who came to Los Alamos. So, anyhow, I was a good student, and that's what we were working on.

4.2.4 Cooper Pairs

When I was working with Bardeen, he first wanted me to apply quantum field theoretical techniques. So I started doing that, in September 1955. I gave some lectures to a small group at University of Illinois on, what were then, the latest techniques such as renormalization methods and functional integrals in quantum field theory. At the same time I was trying to apply them to superconductivity. As I learned more about superconductivity I became increasingly troubled. I remember several lectures by Pippard particularly, where he talked about the simple facts of superconductivity: Specific heat, Meissner effect, etc. It seemed that in spite of the great complexities and differences between metals, there were profound similarities when one entered the superconducting state. Fortunately, we were not aware of or did not focus on the many complexities and variations that have since become evident. But this is the way theories are constructed: First try to capture the new qualitative features. The complications will hopefully fall in place later.

At Illinois we all thought that an energy gap in the single particle excitation spectrum was a key feature that distinguished the superconducting from the normal state. I began to worry that the approach I was taking was too complicated. The problem seemed simple, but baffling. My motto has always been: *Don't try to solve a complicated problem if there is a simpler one you don't understand*. And so I began to think about the fact that the normal metal was reasonably well understood. But as soon as you put in any interaction between the electrons, then you have this tremendous degenerate situation. And nobody understood that. I thought, "That seems really to be a simple problem in quantum theory." I remember talking to Joe Wenescer. "Joe," I asked, "is there some way to solve degeneracy problems in quantum theory?" He said, "Why don't you look it up in Schiff?" I said, "Joe, I already know what's in Schiff, but this is more complicated." So then I went off on a long tangent trying to think about this problem, which really is a problem of a highly degenerate matrix. I got every book I could find on large matrices, stochastic matrices, etc. Bardeen didn't know what I was doing and began to worry about me. But I had a ferocious stubbornness and I just kept working and working. Everybody asks me, "How did you get to the idea of the pair?" What I was trying to do was to find a way to approximate the diagonalization of a very large degenerate matrix. One standard technique is to diagonalize easy sub-matrices first, so I looked for diagonalizable sub-matrices. If we have a pair above the Fermi sea, with zero momentum and spin zero, many states are connected to each other. By then I was making simple approximations, such as neglecting variations in kinetic energy. Then I saw that there were these matrices, these big blocks, so that everything connected to everything. And I could diagonalize them. And as soon as I did that, I saw that you have this coherent state with a binding energy independent of the volume. If you estimated the number of pairs, the entire energy was the right order of magnitude—one could make it a variational solution. Suddenly, there it was; a state that was qualitatively different.

4.2.5 Schrieffer

The key thing was that the existence of the Fermi sea was serving to make the ground state highly degenerate. I was convinced, but to convince everyone else was another matter. I think Bob Schrieffer was the first one that took this seriously. We both saw that the next problem was to put the pairing idea into a wave function for the many electron system.

Bob was there already. We used to talk all the time. He was upstairs, in what they called the Institute for Retarded Studies! He would say, "Bardeen gave me this problem; I'll never get my PhD!"

So we used to socialize together, and ... so, I think between the time I had the idea of the pair, which I think was approximately February '56, and I first presented it in a lecture ... maybe March or April at Illinois—I wasn't as experienced as I am now—what I should have done was to embody the pairs in a wave function so that we could calculate—which is what we were trying to do. But I kept being distracted, proving theorem after theorem, doing Green's functions above Fermi spheres to show that you got bound states and so on. Bob made the next important step; he wrote down the wave function that embodied the pair in a way that was consistent with the exclusion principle. As soon as he did that, we could seriously calculate something. That was what convinced Bardeen. And as Bob said, "We've turned the battleship around." This was about January of '57. I remember that there was some conference that we both went to, and Bob wrote down the wave function, I think he said on a New York City subway.

4.2.6 The Paper

Bardeen was sceptical when I showed him the first paper in letter format. But to give him credit, the letter was sent to him to review. He asked me to change a few things, but then he accepted it, even though he thought it was too abstract, too mathematical. I think it was the end of January of '57, when we started working seriously together. I remember coming in one morning, and Bardeen said, "Let's write a paper together on superconductivity." From that moment we calculated day and night continuously for the next six months, it was the most intense, unbelievable amount of calculations. But you see, we were really ready for it. Bardeen knew every normal metal calculation. And by then I had mastered the electrodynamics. I had been working on that, while Bob had been working mainly on thermodynamics. And we just had to get a few other things into place, the excitation spectrum, how to handle the new matrix elements. There were several wonderful and unexpected discoveries. For example, for a while we would get a London penetration depth too small by a factor of $1/2$. We were tearing our hair out and started to say, "Fine, maybe that's the way it is." And suddenly one time at a concert ... I mean, this calculation was totally in my head ... I saw that you could go from the initial to the final state by two coherent paths. I could do the entire calculation in my head at that time, but since there were several creation and annihilation operators with sign changes, I couldn't be sure of the sign. I could see that the other path would be equal in magnitude, but one sign would give the Meissner effect with exactly the expected penetration depth; the other would give zero. The family fortune was on the roulette table: double or nothing. When I did the calculation, I saw it had come out our way.

I recall coming in early the next Monday morning to tell Bardeen. He listened carefully, as usual, showing no emotion as I was excitedly going through the calculation. When I was finished adding, with a flourish, "So you see, London's penetration depth comes out just as expected," his only comment was "hmmm." But by that afternoon he had modified all of his calculations to include the new term.

Before we solved the problem of superconductivity, everybody considered it extraordinarily difficult, possibly unsolvable: There were theorems that said you can't solve it. But after we solved it, it became regarded as easy. One physicist wrote that it was disappointing that the problem of superconductivity was solved just by this inserting a piddling interaction between electrons.

This is a pattern that occurs in scientific problems. There is a sequence. Before you solve it, people try, but don't succeed. Then they begin to prove that it is not solvable. Finally, someone solves it, and then they say, "Oh, that's trivial, in fact I even ... if you look at the footnote of my paper...." Like the Columbian egg story. If you look at problems that are regarded today as insolvable ... they go through the same thing, and when people solve it, they say it is easy.

4.2.7 Approximations

Another thing, in the way of anecdotes ... one thing I remember is that I was ... after we published the paper of superconductivity..., for a couple of years, in fact, I worked on this with Birger Stölan. I was interested in seeing how you could start from the full many-body problem, and show that in some approximation you would get the kind of pairing that is the basis of the BCS solution. We worked on it for quite a while, and we had enormous numbers of Feynman diagrams. It should have been do-able, but we never really succeeded. I thought that ... since the phenomena occur for an interaction that is infinitesimal, what you could do is that you take a little shell at the Fermi surface; then use a natural small parameter to consider that problem, neglect the variation in kinetic energy... And then you could see that the amount of phase available is a strong function of momentum of the paired electrons. Birger and I worked on this for a long time. We hoped to use this as a parameter in which we could make an expansion, and show that somehow the pairing came out; it turns out that it's not easy. Anyhow, I was giving a lecture on this at some point. And after the lecture a young man comes up and says: "I don't understand why you are working on this problem, everybody knows that there is a pair condensation." And this was only a couple of years after I was a voice in the darkness myself, so times change.

In the expression for T_c, we introduce the Debye frequency ω_d. This is quite a high frequency, and someone may ask why such a high frequency is relevant at very low temperatures. The reason for using that, is that was the range, according to a simple field-theoretical calculation for which the interaction would be attractive. And you see, the intriguing thing was that you didn't use most of this range in the weak coupling approximation because of the exponential factor.

4.2.8 Breaking the News

Among the people in Illinois we had lunch every day and told each other what we were doing; so they were terribly excited. And then what happened was that we submitted a letter and then two post-deadline abstracts to the March meeting. Bardeen wanted Bob and me to present it, which was very generous. Then what happened, was that I got to the meeting, and Bob went to New England to visit a friend, and he couldn't make it to the meeting. I was carrying his slides. So I ended up giving both parts of the talk, both his and mine. The reaction at that meeting was fantastic! There was a big audience, because people had already heard... Packed chamber. Fantastic reactions. I didn't appreciate how fantastic it was. I don't remember everything. Some things take time, but BCS gained instant recognition, instant. But then there were complaints about gauge invariance. Our attitude about things like that was that, "We'll clean it up afterwards." The big fuss was that for the theory to be explicitly gauge invariant we had to include longitudinal excitations ... but we knew that longitudinal excitations, because of the long-range Coulomb forces, would be of very high energy. Bardeen and I used to talk about it, and then we just said... "Sure, we have longitudinal excitations, but we don't have to worry about them."

Various people attacked this question, Anderson and others, I think it was solved best by the Green's function methods. There were various things we did, that had to be cleaned up. A new set of ideas often has initial inconsistencies (consider the Bohr atom). You can have two attitudes: one is that it is inconsistent, so throw it away, and the other is that there is so much in the ideas, the problem is to make them consistent. It's like a little boy who is an optimist, so optimistic that his parents wanted to teach him what the world is really like. And so for Christmas they gave him a pile of manure. They tiptoed down Christmas morning, and they hear him singing and whistling, singing and whistling, and they went to look in his room. And there he had a shovel, and he says: "In all this manure, there must be a pony some place."

Consider Dirac's vacuum. With the stroke of a pen, he changed the concept of vacuum from the vacuum as a void returning to the Greek notion of the plenea, you know the problems: electrons interacting with each other with infinite charge, etc. But the thing to do is to really make it sensible, this happens over and over again.

It was something we worried about: I remember that I talked to Bardeen about it, we knew that there would be longitudinal excitations; they had to be there. But they would be at a higher energy; that's the current view now. You have gauge type theories, and massive bosons. So it didn't particularly surprise me, I guess... but everyone else might ask, "Why didn't you work on it?" Well, frankly, I was exhausted. Just exhausted. This problem was understood very quickly. The big problem for the Nobel Prize was that Bardeen had already won the Nobel Prize in Physics. I think he is the only person to have ever received the Nobel Prize twice in the same field. (There may be another in chemistry.) That was a problem. I was told the committee thought for a long time if they should have some other combination. But, frankly, I think that they did exactly the right thing. It would have been terrible to leave Bardeen out; he felt that superconductivity was so much more important than the transistor as a scientific problem. Superconductivity was the holy grail. So I think the committee did exactly the right thing, but I believe that was what held them up. That's why there was a long wait. After the pairing idea was conceived, I was convinced that this was the solution; I was terrified, but I was convinced. It was a very painful period; because it was very difficult to convince others, and there were some very famous people out there.

Well, "knew" is a strong word, but I certainly believed that the pairs were the basis of a solution. But after BCS there was no longer any doubt.

4.2.9 Contributions from Other Scientists

There were other people who contributed in this process, that helped during the period of calculation, Charlie Slichter in particular. The atmosphere in Illinois was that everybody helped everybody. We talked at lunch almost every day, as I said. Bardeen was interesting in this respect, because with all his talent, he did not have a great capacity to communicate. I remember that we'd spend all morning defining symbols, terms and kinds of stuff. Then we got to lunch, and he would start to talk to the people as if they had been there in the morning conversation. You could see that they were really puzzled. But we'd talk and talk, so there might very well have been suggestions. Charlie Slichter is one such person. He learned how to do the calculations, and was doing them as fast as we were. I remember he found some errors.

His experimental work on NMR is also important. Later, flux-quantization was demonstrated, and we of course learned about it when it happened.

And then the Josephson junction. We were so close to having the Josephson function. Because actually at that time I was working on the proximity effects; the penetration of the wave function... But then, you can't win them all.

The controversy between Josephson and Bardeen I do remember discussing with Bardeen. I believed in Josephson's result, but Bardeen had his doubts. I don't think he really thought of the wave function going coherently into the junction. I'm not sure about his reasons. But I remember talking to him about it. I believed it, because I had been working on coherence. If he didn't believe it at first, he came to believe it eventually.

4.2.10 Present Problems (2001)

I have long ago switched to biology, working on the interaction of many cells instead of the interaction of many electrons. That was one of the things that attracted me in the first place.

I think that if I have any gift, it is to look at very complicated problems and try to extract some simple essence that contains the qualitative effect. That's really the nature of good physics, very often. When you look at for instance ... a very good example is phase transitions, critical phenomena, you literally do this, you throw something away, and you know exactly what you throw away. And you only keep the essential part, and that does everything. That seems to me to be the essence of really good physics. You know, Einstein said, "Make things as simple as possible, but no simpler." You have to take an intellectual risk, because what you throw away might be the essence. And you also take a political risk, because what you throw away might be something that someone worked on for their entire life. And especially in biological problems, you get some very aggravated people. What I focused on is: What is the molecular and cellular basis for learning memory storage? Where and what takes place on a cellular level, when we learn? And at first it seemed like an impossible problem, but nowadays I think we've almost solved it.

We think we know the systematics of the changes, how the changes depend on receptors and molecules involved and variables such as cell firing rates. One of our earliest formulations—25 years old already—is known as the BCM-theory. It stands for three authors, Bienenstock, Cooper and Munro. Most people in neuroscience know me by BCM. Once in Washington someone was introducing me and mentioned BCS. The woman sitting next to me, a very famous neurophysiologist, said: "Didn't he get that wrong?" And I said, "No, no, it's ok, don't worry about it."

So I like to say that I have the good fortune of being sandwiched between very gifted colleagues!

BCM's postulates have also been tested experimentally, and confirmed. And some of the very subtle consequences have been tested experimentally. Experiments have been designed expressly to test the consequences of this theory, and they have uncovered new phenomena. It is very rare that this has happened in neuroscience. And to me, if we do this, this will be one of the great achievements ... to bring serious mathematical structure into neurophysiology. After all, that is what Galileo did for physics, isn't it? We are now at the stage where we are working on the underlying molecular mechanisms. In fact, I am just working on a paper with some colleagues, which proposes a single underlying molecular mechanism to explain synaptic modifications that are known to occur. And we know that there are changes that take place in certain molecules, and in the cell-surface there are so many things going on ... so much though, that I would say that this particular problem of cellular basis for learning memory storage probably will be resolved in the next couple of years. The next problem is how this is put together when we do processing. And the hardest problem is how you construct mental states, how we become aware of ourselves, conscious and so forth. And these problems are regarded as so difficult, that as you know, some people say they are insolvable.

Whether it is solvable you never know until you solve it. But ... why shouldn't you try? After all, a hundred years ago people would have said that there is a fundamental distinction between living and unliving. Between organic and inorganic. And then a hundred years before that, between celestial and Earthly material. And now, not only have all of these things been resolved, but we see that the distinctions aren't very good. That there is no clear boundary between living and unliving, or whatever you want. It is a matter of definition. And the same things might be for mental states, we don't know. But then again you don't know until you've solved it, but when you've done that, it looks easy.

4.2.11 In Between?

I think I would classify a virus, not prions, as more clearly something in between. Something that can reproduce when it captures cellular machinery, but which, without that cellular machinery can't do the job by themselves. It seems more likely that a prion is a protein gone bad. In fact, when you think about it, it is somewhat miraculous that with all the proteins floating around, they don't get in each others way very much. Well, of course if they got in each others way things wouldn't succeed, so part of the process of evolution is to sort that out. But you can ask conceptually, suppose there is a protein that interacts with another protein in a devastating way;

that would be highly poisonous. And if it can interact with the protein in such a way that makes the protein make a transition from its initial folding to another folding, which then makes the prion duplicate itself, then you would have a very deadly possibility. We have been thinking about this a lot—trying to find a model. But I don't think that really should be classified as a living thing. You know, that's just a question of definitions. You can say that a living thing is something that can reproduce, you can say that a living thing is something that can speak, or ... I was once having a discussion at a dinner in London with a lady. I said, "It's a continuum, because you can teach apes language." And then she says: "Oh no, an ape can't quote a line of Shakespeare." And I thought to myself that there are many human beings who can't quote a line of Shakespeare.

So you can make the distinction as you wish, but my own opinion is that it's a continuum. And maybe that goes for consciousness. That the transition between conscious and unconscious in phase transition language is second order—consciousness comes out of the shadows and we gradually become conscious. This might very well be the case for every human baby. But the heart of the problem, as I like to put it, to paraphrase Santayana who once said: "All of our sorrow is real, but the atoms of which we are made, are indifferent." To my mind, the problem can be stated very simply: How do you make real sorrow out of hypothetical atoms? That's an unsolved problem. And that is the problem that some people think is unsolvable. Their use of words such as "consciousness" is subjective. My answer is: give us a couple of years. Everything is unsolvable, every hard problem has been unsolvable, but then it is solved and then it is trivial. It's like what we talked about just a minute ago. Well, that doesn't mean that we will always find solutions in the usual way. I can't prove that you can find solutions, but personally I think that throwing up your hands and saying you need another law of nature and so on ... why give up so fast? But then you see here I am just a card-carrying, working day scientist. A "no nonsense physicist," as one reviewer of my elementary textbook wrote some time ago. That's what we all believe.

Let me close my story here by referring back to Stockholm, December 11, 1987, the day after Alex Müller and Georg Bednorz received the Nobel Prize for the discovery of high-T_c superconductors. The laureates came down the stairs in the physics auditorium at KTH, and there was a tremendous applause, even standing ovation as far as I can recall, from a packed auditorium, a nice moment. I was sitting in the panel, and was challenged to express my views. I talked about the possibility of the new high T_c superconductors going from pairs with very large coherence distance near T_c, to those with a coherence distance so small they are like real bosons. This might not be possible in the present high temperature superconductors, but recently, in the Bose-Einstein systems with tuneable Feshbach resonances one seems actually to see the transition between coherent BCS pairs and bosons.

Chapter 5
John Robert Schrieffer: The Microscopic Theory of Superconductivity

"I happened to be in New York, and I was sitting on the subway when I realized that maybe the scheme that Tomonaga used to describe the pion-nucleon interaction would be a useful way to go."

Fig. 5.1 John Robert Schrieffer

5.1 Biographical Notes

John Robert Schrieffer shared the Nobel Prize in Physics 1972 with John Bardeen and Leon N. Cooper *"for their jointly developed theory of superconductivity, usually called the BCS-theory."*

John Robert (Bob) Schrieffer was born in Illinois in 1931. In 1940 the family moved to New York, and later, in 1947, to Eustis, Florida where the family engaged in the citrus industry. Schrieffer started on an electrical engineering education at MIT in 1949. His interest in this field came from personal experience as a radio amateur on a homemade "ham" radio in young teenage years. But at MIT he discovered, through his own reading, the challenges and fascinations of physics, and made the switch to physics after two years. Under John C. Slater he did his bachelor's thesis on the structure of heavy atoms in 1953.

He became interested in solid state physics, and began graduate studies with John Bardeen at University of Illinois. He did research, both theoretical and experimental, on semiconductors the first two years. In his third year, on the advice of Bardeen, he started collaboration with Leon Cooper, and the three together were committed to solving the superconductivity problem. While the three men struggled with that problem, the young PhD student Robert Schrieffer felt uneasy about progress, and without telling his adviser, professor Bardeen, he conducted a separate research project in ferromagnetism as a safeguard against a possible failure to solve the superconductivity problem. When Bardeen was about to go away for a meeting in December 1956, he suggested Schrieffer should go on working on the superconductivity problem for yet another month before changing subject, because he felt they might be able to solve the problem. While Bardeen was away, Schrieffer happened to be on a visit to New York. Sitting on the subway he realized that the Tomonaga approach might be the way to do it in a consistent way. He wrote down the wave function, now known as the BCS wave function, and calculated the energy of the system. It had the same form as the Cooper solution, but was exponentially stronger. On his return he told Cooper, and then Bardeen, who agreed. During the next 11 days they worked out the thermodynamics and other properties. Their analysis agreed with experiment.

The paper was published in Physical Review in 1957, known as the famous "BCS-paper." Schrieffer emphasises that the BCS-theory has a much wider validity than just the phonon mechanism, referring to the applicability of the BCS-theory in totally different systems like in nuclear matter and in neutron stars. Robert Schrieffer has had a distinguished career. He spent the first couple of years after his thesis work on the BCS-theory at University of Birmingham and at the Niels Bohr Institute in Copenhagen, and then at University of Chicago and University of Illinois. In 1962 he joined the faculty of the University of Pennsylvania as professor. From 1980 to 1991 he was a professor at University of California in Santa Barbara, where he served as Director of the Institute of Theoretical Physics from 1984 to 1989. He was later called on to become University Eminent Scholar Professor in Physics at Florida State University in Tallahassee, Florida, and was Chief Scientist at the new National High Magnetic Field Laboratory (NHMFL) since 1992. Professor Schrieffer holds seven honorary doctorates, and has received a number of prestigious awards, foremost among these the Oliver E. Buckley Prize from the American Physical Society in 1968, and the Comstock Prize from the National Academy of Science 1968. He received the Nobel Prize in physics for 1972, shared with Bardeen and Cooper, for the theory of superconductivity. Robert Schrieffer was Vice President, President Elect, and President of the American Physical Society 1994–1996. His profound insights and his kindness to colleagues and friends has been greatly appreciated in the physics community.

As for the impact of superconductivity science in general, we quote here from Frank Wilczek—as conveyed to the present author—some statements Schrieffer made to him: "Understanding superconductivity either initiated or brought to a new level three big concepts in foundational physics: gauge symmetry breaking ("Higgs mechanism"), pairing, and field topology. Gauge symmetry breaking underlies the modern theory of electroweak interactions. It also enables us to formulate more ambitious unified theories that bring in QCD as well. Pairing ideas, applied to quarks and antiquarks, give us an amazing and successful theory of pi mesons. Putting pairing and gauge symmetry breaking ideas together, leads us to a truly elegant theory of ultradense hadronic matter, which explains quark confinement and mass generation in that regime analytically. And superconductivity also makes a very direct contribution to particle physics: Running a machine like the LHC would be prohibitive, economically, if its magnets weren't superconducting!"

5.2 His Own Story

5.2.1 Early Inspiration

Regarding my background, my family were in business; my father was in pharmaceuticals and cosmetics. And so he didn't really have any direct scientific background, himself. I was very interested in chemistry at one point and I had a friend next door who joined me in making some rockets. So we had great fun in launching these rockets, and the family was very concerned about the safety of the entire operation. But we did have some fun, and no-one was injured, even though our rockets went up about 150 ft. They were made out of cardboard tubing and the powder was put into that, and that acted as a small rocket. I think that young people at that time were interested in having chemicals, hence they could perform experiments, and I did the same.

Later, after that I became interested in ham radio. It's an amateur radio where you can transmit and talk with people around the world. I was a babysitter and the gentleman whose house I sat in was an amateur radio operator. He had some books that he was interested in. I read those books and got very excited about having a ham radio myself. I got to Switzerland and Hawaii and all over the United States. I was about 15 at that time. I was a registered radio amateur operator. We sent postcards to each other all over. My call number was w4oay, "w4-over-anxious-yankee." I had great fun with it. I ended that career when I went off to college at MIT.

At the time I came to MIT, I was planning on becoming an electrical engineer. So I began studying engineering, and I did that for two years. But I picked up a book on atomic physics by Max Born. And in the back there were about 24 appendices, and each one talked about some aspect of quantum mechanics. So I got very excited about the physics aspects, and in the end of my second year at MIT I transferred to become a physics major. And I was really excited about physics at that point. This was in 1949 through 1953 when I was an undergraduate at MIT.

5.2.2 At MIT

At MIT there was a lot of competence in radar technology. But my interest in electrical engineering came from my radio amateur experience. My thought was, I was interested in continuing along that line because I had much fun with that. However, while I enjoyed electrical engineering I found physics much more fascinating and a deeper subject. So I decided to go in the physics direction for that reason.

I really enjoyed MIT, it was a wonderful, stimulating place where you could go to the laboratories and talk with the people doing forefront research. And I did a senior thesis on the magnetohydrodynamic effect. We made a coil, a magnetic field, and had a small wire coming down into a bottle. And the bottle had inside of it a floater which could tell the orientation of the liquid, as it moved. At the bottom of the bottle we had a rotor which would cause torsional excitations, so when the magnetic field was on, it would cause theses torsional modes to move up the column of mercury. So we could measure that by torsion wire. That was the first experiment we tried without any background from the lead professor. We observed these torsional modes, both the frequency and the amplitude of them.

The professors were very impressed, and they had to get down on their stomachs to look at the actual experiment, since it was underneath a laboratory table. We had a lot of fun with that. We should have published it, but the professor did not tell us to do it, so we just wrote up the thesis.

5.2.3 Ambitions

I was interested more in the fundamental aspects of physics, rather than the applied aspect of it. What I wanted, was to become a researcher in physics. I thought becoming a professor would be a good way of doing that. So, I pursued physics hoping to go on to graduate school, and then, perhaps, become a faculty member at some university and carry on research.

Originally I was interested in going into nuclear and particle physics, and I had a Fullbright scholarship to go off to University of Birmingham. And what I did then was to make plans to head off. But then it was the Korean wartime, and my parents were worried I would have problems getting back from Europe, which certainly wasn't the case. But John Bardeen offered me a position where I could start with research immediately by going to the University of Illinois. And I was knowledgeable that he had been working on superconductivity in the past. I thought that would be a marvellous topic to be involved in. So I went to the University of Illinois, but he originally sat me up looking at some problems in semiconductor surface physics. It was the problem of electrons scattering from a surface and what the mobility would be for electrons parallel to the surface when they scattered off the surface as they moved. And I worked out a theory of that when there was an inversion layer of the potential coming near the surface. And what we did was to then calculate how the electrons scattered off this potential barrier and also off the surface. And we calculated the effective mobility you would get. So that was during my first year at the University of Illinois.

5.2.4 First Publication

I published that, and one thing I predicted was that the levels in the well would be quantized. And that led to the whole thing with the hole-effect, if you like, for quantization in the z-direction. This was the first in that area, and I was very proud of it. Others followed up this work later. Then I went to the semiconductor laboratory with Bardeen, to do some experiments to prove that the theory was ok. We managed that. So it was an exciting time also. I really considered myself a theorist, but to go into the lab I thought was a good possibility of broadening oneself. I would think this would be possible even now. In the beginning of a project a student could go off and do some experiments for a year, either on a theory or some other topic, and then come back. I should remark that this was an interesting warm-up problem that took two years.

At the end of the two years I came and asked for a thesis problem. And Bardeen had in his drawer a list of ten problems. One of them was the resistance minimum, which occurs at low temperatures when there are magnetic impurities in metals, later called the Kondo effect. There were eight other problems, plus superconductivity. So I said that I would like to work on superconductivity.

I went to my professor, Francis Low, who was a particle physicist, and I asked him what he thought about working on that problem. He said, "Well, how old are you?" I said I was 23. He said, "Well, that would be a good problem because you're young enough, and if you fail you can do another problem." So I picked that one. By that time Bardeen didn't yet have the grand plan to solve the problem. He had basically not chosen a problem at that point, but he suggested that I look at the problem of some mathematics which had been developed in nuclear physics, by Bruckner and others, which was involved with a scattering matrix approach. And that's what I developed in the thesis. And I found that there was instability in the scattering matrix approach which corresponded to pairs of electrons interacting. But Leon Cooper had been studying, as you know, two electrons interacting above a Fermi surface that was quenched, and found the bound state. Well, I found the instability in the many-body problem.

5.2.5 Thinking Big

Regarding the question that triggered Bardeen to think big, I think he felt that this instability which Leon had seen in the pair problem would be an important thing to follow. And if one pair was instable, it was clear that in another region of space other pairs would be unstable, and the instability problem was how these pairs interacted with each other.

So when he hired Cooper he hadn't yet thought of this. He thought that Cooper would be able to bring some methods from quantum field theory to bear on the problem of superconductivity. But the problems in quantum field theory at that time were perturbation problems in electrodynamics. And it was clear from the theorem of Shaffroth that superconductivity could not occur in any power of the interaction. It had to be a singular form of the coupling in order to produce superconductivity. Fröhlich was going in the right direction, but he ran into mathematical difficulties which made it impossible to treat the many-body problem in a consistent way. And Bardeen also developed a theory with small energy gaps, and that ran into mathematical difficulties as well. So we felt that those directions were not the correct ones to go.

When the three of us got to work together, it was a declared purpose by Bardeen that the three of us would solve the big problem. It wasn't clear how to go about it, however. I had very good hopes myself that this would work out, because I felt that Bardeen was a brilliant man and that we had a good chance of making some progress. Of course I didn't have the background to judge that realistically. But Bardeen had been working on the problem since before the Second World War, and he was interested in making a new push at that time.

So, we were interested in pushing the problem, but it was not clear whether we could solve it or not. Feynman was working with superconductivity in Japan at that time, and that was one of the pressures we felt; that we really must work hard to be able to beat Feynman, because he had done such beautiful work in superfluidity of helium. And Feynman was working on the specific heat looking for the anomaly in the second order phase transition. But again, using perturbation-theory, and that simply could not work.

Regarding Lars Onsager, we were not particularly concerned about him, because we didn't think he was working on the problem. He had made some guess in the direction of pairing, but without any theory behind it.

5.2.6 Success!

To summarize the recipe for the success of the BCS team:

First was Leon's discovery of the instability of the system from a single pair point of view, and that was very important. The next was the writing down of the pairing Hamiltonian which would describe the system of zero momentum pairs, and the total momentum of each of the pairs was the same. Bardeen felt that the best way to describe the problem was a superposition of quasi-particles, because the condensation energy was only one part in 10^8 of the total correlation energy. He felt that it was really important to be able to isolate that paring problem, which is the singular part, and let the rest of the problem be described by the Landau theory of the Fermi liquid. So that was the basis for the pairing approximation. We felt that treating the zero momentum pairs first would capture the most important part of the problem. And after that we could treat finite momentum pairs separately. Briefly stated, the method was that of pulling out the critical part of the problem, and letting the rest of it be treated as a perturbation. This was the method we chose.

To complete the story I should include the following: I had decided to work on ferromagnetism quietly, without Bardeen knowing about it. And I started doing that in parallel with superconductivity. And then Bardeen went off to a meeting in December of 1956, and he said; spend one more month on this before you change, because I think there is a chance that we can solve it. And while he was gone, I happened to be in New York, and I was sitting on the subway when I realized that maybe the scheme that Tomonaga used to describe the pion nucleon interaction would be a useful way to go. So we were interested in treating the pairs in this consistent way. And I wrote down a wave function which is now called the BCS wave function and calculated the energy of the system, and found it was singular, and it had the same form as Cooper had found, but was exponentially stronger than the Cooper bound state.

So I felt that this was an interesting development. I came back to Urbana Champaign, where I met Cooper at the airport and told him about it. And the next day we both went in and told Bardeen. After he had looked at the problem, he said: "I think that's the answer. That solved it!" So we quickly—in about eleven days—calculated the thermodynamic properties and the electrodynamic properties, and saw that it really did agree with experiments. The first problem was calculating the condensation energy in terms of the gap. And Tinkham and Glover had done some experiments which showed the size of the gap in infrared absorption. And from the relationship between the gap size and the condensation energy, we were able to see that they were consistent, and it proved that the fundamentals of the theory were correct.

5.2.7 Ultrasound

At that time there were also ultrasonic experiments going on at Brown University, by Morse and coworkers. The so called coherence factors which separated out acoustic attenuation from the electromagnetic absorption were very important. And extremely important were the experiments of Hebel and Slichter in the nuclear magnetic resonance, which showed that the relaxation rate went through a maximum before dropping down to a low value at low temperature. That proved the existence of the coherence factors, which were characteristic of the pairing state. And that the acoustic attenuation just dropped continuously below T_c, which was another prediction of coherence factors. So that was another strong indication that the theory was correct. It's when you excite across the gap it has the reverse coherence factor. If you make a very high frequency acoustic attenuation across the gap, the coherence factors go the other way, and you see a peak rather than this sudden drop. There were also specific heat data around at that time, which supported our results, and another feature was the isotope effect, which came out in a very natural way. The isotope effect led us specifically to guess that the mechanism was a phonon mechanism.

5.2.8 Transitions

Some people are puzzled by the fact that the characteristic frequency in the problem is the Debye frequency, which is a high phonon frequency, while the physics happens at low temperatures, which would seem to call for a low characteristic frequency. One has to bear in mind that the transitions involved are virtual transitions rather than actual transitions, and the virtual transitions went up to high energy just because the interactions were fairly strong. It didn't have to do with the physical temperature. Even at zero temperature, where you don't have any excited quasiparticles the excitations went up to the Debye temperature where interactions change sign and become repulsive. So in order to get a bound state you had to use all the attraction you could find. That occurred all the way up to the Debye temperature, the Debye frequency.

One problem was that we didn't have the right treatment of the pairs that were doubly occupied by the excited quasi particles. So we didn't really have the specific heat done correctly at that time. Then we published this letter, which explained the NMR and the acoustic attenuation, and we didn't have the second order phase transition at that time. We found that we had to orthogonalize the excited pairs of quasiparticles to the ground state, which we hadn't done correctly. And all of a sudden the second order phase transition came up. So that was the remaining piece of the puzzle that we reported in a long paper.

Gauge invariance was a problem people on the outside were worried about. We chose a particular gauge because it was a simple one to calculate, and we felt that any other gauge would also give the right answer. But one had to be careful about how you treat the collective excitations in the system. And if you included the collective modes, that would lead to a manifestly gauge invariant theory. And that's what turned out to be the case. We knew from our understanding of the problem that the gauge problem would be solved.

The time spent on working out what became known as the BCS-theory, was a brief and hectic period. It lasted from January through about April 1957. We worked hard because we were very enthusiastic, but we were also worried about Feynman working on the problem. And then Schafroth was also working on the problem. But we were convinced that they were going in the wrong direction, they were considering the dilute pair case, rather than the strong pair case. And they were considering the interactions between pairs basically in a perturbation approximation, and we were convinced it had to be a non-perturbed interaction, the way the BCS theory worked out. People have compared it to the competition to solve the DNA-problem. Pauling was considered by Crick and Watson to be a strong competition, thus motivating them to work very hard. But I think we were mainly motivated by our excitement to work through this problem, even though we had Feynman breathing down our neck.

5.2.9 Finished Work

At the end of this period we were feeling really marvellous, elated, we were really excited about the whole thing. And we were convinced the theory was correct. We published in Physical Review later, and before that we submitted a post-deadline paper to the American Physical Society meeting, in which Leon Cooper and I were recommended by Bardeen to give a post-deadline paper, which was remarkable because he let the young people go forward and announce the theory rather than himself. He was a very selfless person in that regard.

So we presented this paper at the March meeting, and that was the first announcement of the theory at that time. Unfortunately, I could not make it to the meeting because I got stuck in the snow in Vermont, so Leon gave the paper. And although I lost some of the fun, he gave a good description of everything when we met.

I think the world reacted in two ways; those who believed that all this experimental confirmation of the theory really showed that the theory was correct, while others who looked at it felt there were problems. For example Henshow felt that the problem of lack of manifest gauge invariance showed that the theory was not really consistent. And Bloch also felt the same thing.

I think for a year or two there were concerns about it. But then there were proofs that if you included the collective modes in a consistent way, the theory was in fact gauge invariant. That was done by Anderson and by Rickayson, and there were also many others. Nambu also had a theory, which showed that it was gauge invariant, if you included the collective modes. That removed all the problems.

Afterwards, I thought of going to Berkeley, and at Caltech there were possibilities to come into the faculty. But I went to the University of Chicago for one year, and then joined the faculty of University of Illinois, where I spent two years. Bardeen and I were going to write a book together on superconductivity. Instead we had a very long chapter for a book put together by Gorter, and decided not to write a book about superconductivity at that time. When I went to the University of Pennsylvania I taught a course on superconductivity, and I decided to write a book.

The work on the BCS theory was joy and pleasure, because when it was worked out the theory also applied to nuclei. Aage Bohr and Ben Mottelson had picked up the idea of pairing. There was already an even-odd artifact effect in nuclei which showed that the energy to add a particle to the nucleus depended upon whether there was an even number or odd number of nucleons. But there was no clear understanding of how to treat the many-body effects associated with the pairs. And when the BCS theory came along, it really explained that. They received the Nobel Prize with Leo Rainwater for the collective modes in nuclei, and part of it was the pairing, but there were many other features associated with their work.

5.2.10 Josephson

Later came the Josephson equations with the Josephson predictions. There is this story, which is true, that Bardeen was sitting in the first row listening to Josephson's talk, and he didn't believe it. He felt that the problem was that there would be extremely weak coupling because as one went through the barrier, the gap was basically going to zero, and that the coupling between the two sides would be exponentially small, due to the fact that the pairing wave function was decreasing inside the barrier to an extremely small value.

He was wrong because the size of the coherence length is large compared to the thickness of the barrier, and therefore you shouldn't take a local view of how the pairs were being decoupled but rather take a more global view of it. And it was this fact, that the pairs had a large size and could tunnel coherently from one side to another with a large gap, that leads to the effect. He then quickly realized that his point of view was not correct, and he was very much interested in the Josephson effect, and he published on it. At that point I had already left.

People have often asked why it took so long to get the Nobel Prize. I don't really know. Part of it was the fact that the theory didn't agree in all numerical details with all superconductors. There were the strong coupling effects which made lead have somewhat different characteristics, off by about 10 or 15 percent from the theory. And there were lurking suspicion, I think, by a couple of people in the Swedish Academy that the theory was not completely correct, because of these small deviations.

But then the strong coupling effects were treated by the Eliashberg formulation of the theory, which showed why the deviations were true, and had just to do with the fact that there was quasiparticle damping, and the retardation effects were also important. So we had to include those effects in order to make the theory quantitatively in agreement with all the experiments.

I think Abrikosov, and Gorkov also, who derived the form of the Ginsburg-Landau theory from the BCS, both deserve the Nobel Prize. We were aware of it, of course, and felt that the important features of it were the fact that there was a condensate, and the condensate would lead to the order parameter, which was in the Ginzburg-Landau theory. But we didn't have a derivation for it, and that's what Gorkov provided, in a very beautiful way. It was published in 1956.

It is surprising that Abrikosov didn't publish his work regardless of what Landau said. But I'm sure that Landau was a very influential person, it was difficult to move ahead without his blessing.

I had no personal intention of working on the relationship between the BCS and Landau theory, and I'm not sure if Bardeen was interested in it. Cooper left the field pretty much after he completed the work with us. So he was not that particularly involved there. But I would think that—I certainly didn't—maybe Bardeen had thought about the problem. But I don't think he published on it.

5.2.11 Rating Superconductivity

Superconductivity has had many, many surprises in it. And the theory is very widely applicable, even in particle physics where the condensate of the cores is basically like superconductivity; it's a pairing condensate. So that both nuclear physics, and the neutron stars are nice examples of BCS-theory applied to the condensate of neutrons. So I think it's a very broad theory that broke new ground, in it not being analytic. It's also widely applicable to fermions in any system.

I believe that the high-T_c is still a very challenging problem, and the final resolution of that from a theoretical point of view is still ahead. The materials are much more complicated than we were considering, and the correlations being strong lead to important effects in the normal state, while we could use the Landau theory, and basically not worry about the complications of the normal state. That is certainly not the case for the high-T_c materials, where normal state correlations are very important.

I think it will be solved, but it probably will not be solved in any simple way. There are aspects in the problem which, I think, are brought up by the fact that even computational studies have not been able to prove that superconductivity occurs in any simple way. It's a very naughty problem by requiring understanding the normal state first, and we don't really know what's going on there terribly well.

We didn't have any direct feelings about whether we would ever see superconductivity above 77 Kelvin. I think about 22 degrees Kelvin was a simple effect. We didn't know if this was an experimentally limiting effect or not. There were so-called proofs that the electron-phonon interactions could not produce a T_c above about 35 degrees. But we didn't think that was the only mechanism, there could be other mechanisms that could produce a stronger coupling and make the T_c higher, which turned out to be the case in the end. We had no direct predictions about that, except that the formula for T_c had this characteristic $N_0(V)$ in it, and if that is very large, and also the pre-factor instead of the Debye frequency was a much larger frequency due to another mechanism, you could get a higher T_c.

I believe that we will reach room temperature superconductivity, and it has to do with whether or not you can find interactions even stronger, even the spin fluctuations, which leads to very high T_c, as we know. And perhaps if you found something very strong as exchange coupling, which would lead to a stronger coupling between the excitations, one would be able to form a higher T_c material.

I am often asked why I moved from beautiful Santa Barbara to Tallahassee. The answer is that the National High Magnetic Field Lab had a unique possibility of forming a group that we felt would lead to important contributions in all areas of condensed matter physics. It was a new venture that I found very exciting. I'm delighted to be here. Also, I spent two years of high school in Florida, but it was my parents' home state, basically, not mine. Some have viewed this move as a sacrifice. I do not; rather I see it as a great opportunity, and that has turned out to be the case, so I'm delighted to be here.

Chapter 6
Ivar Giaever: Single Particle Tunnelling: Confirming the BCS-Theory

"After I got a "go-ahead and try it," it took less than a week before I had done it, even though I had never done superconductors before."

Fig. 6.1 Ivar Giaever

6.1 Biographical Notes

Ivar Giaever shared the Nobel Prize in Physics with Leo Esaki and Brian D. Josephson. More precisely: The Nobel Prize in Physics 1973 was divided, one half jointly to Leo Esaki and Ivar Giaever *"for their experimental discoveries regarding tunnelling phenomena in semiconductors and superconductors, respectively,"* and the other half to Brian David Josephson *"for his theoretical predictions of the properties of a supercurrent through a tunnel barrier, in particular those phenomena which are generally known as the Josephson effects."*

Ivar Giaever was born in 1929, and grew up in Toten, Norway. His career in physics is an unusual one.[1] After high school, his priority was to study electrical engineering at Norway's leading engineering school, the Norwegian Institute of Technology, in Trondheim, since 1996 incorporated into the Norwegian University of Science and Technology. However, the competition to get in was very strong,

---

[1] See *Kristian Fossheim: Ivar Giaever* in *Norwegian Nobel Prize Laureates*. Olav Njølstad ed. Universitetsforlaget 2006.

and due to the equivalent of a C in a Norwegian language course and in French, he was not admitted to study in the Department of Electrical Engineering, but had to accept mechanical engineering, a subject in which he took no real interest. Consequently, by his own account, he spent his student years in Trondheim from 1948 to 1952 mostly doing other things than study, specially attending activities in the lively Student Society.

In postwar Norway, after the Nazi occupation, the housing situation in the cities was extremely difficult. Having found a job in the Norwegian Industrial Property Office in Oslo turned out not to be of much help. The young family decided to emigrate to Canada in 1954. Giaever joined the Canadian General Electric where he enrolled in their advanced engineering program. Soon, the family moved south of the border, and Giaever joined the General Electric Company in Schenectady, New York, where he, now also as a serious and hard working student, completed the company's demanding engineering courses. For a while he worked as an applied mathematician on various assignments.

Giaever felt greatly attracted by the opportunity to do research within the company, with its impressive staff of skilled scientists at the GE Research and Development Center. He joined the center in 1958, and concurrently started to study physics at Rensselaer Polytechnical Institute (RPI), where he later was to earn his PhD in theoretical physics in 1964. It was during class in a physics course in 1960, where superconductivity was being taught, that the idea struck him how to detect the superconducting energy gap which the BCS-theory had predicted. He realized such a gap would affect tunnelling characteristics for electron transport through a barrier between thin metal films. All experimental facilities he needed were around, along with the support of highly trained scientists. Giaever made his thin film structure, and could soon, in 1960, "measure the energy gap in a superconductor with a voltmeter," as he later put it.

Against the background sketched above, this became next to a scientific sensation in the physics community. The mechanical engineer, now a physics student, had done an experiment the greatest experts could only have wished to do, but did not conceive. Adding to this the great ability Giaever has to communicate his work in lectures, has made him an attraction at meetings, at universities and research institutes. The fact that he is always open and candid about his unusual background as a physicist, has added a special flavour to his story and his work. His follow-up research on the density of states near the superconducting gap, matching very closely the BCS prediction, demonstrated beyond any doubt that his discovery was no accident. It also made very clear that his original observations of tunnelling between normal metals had been correct. The Nobel Committee considered his verification of the BCS-theory to be his main achievement.

Giaever insists that some element of luck was involved, and comments that this is needed to succeed. One should not be tempted to think his success came easy, however. Many years of hard work at GE was behind it all. Ivar Giaever continued his tunnelling experiments for several years. It may also have been an inspiration for Josephson's discoveries. Giaever was also first to publish measurements showing a finite current between two superconductors under zero voltage condition, what later became known as the DC Josephson effect. However, due to the fact that the whole idea of a zero voltage supercurrent across a barrier between two superconductors had not yet been formulated, Giaever never laid claims on having discovered the Josephson effect.

When Ivar Giaever received the Nobel Prize in physics in 1973 he had already left superconductivity, and had started work in biophysics during a stay at Cambridge, England in 1969. His special area has been the behaviour of protein molecules at solid interfaces. He left General Electric in 1988 to become an Institute Professor at Rennselaer Polytechnic Institute in Troy, and has for a number of years, concurrently, been a professor at the University of Oslo, Norway. He is the recipient of numerous honorary degrees and prestigious prizes, among them the Oliver E. Buckley prize.

6.2 His Own Story

6.2.1 Background

I grew up in a place called Lena, in Toten north of Oslo, which everybody in Norway knows is farm country. My father worked in a drugstore there, belonging to my grandfather.

I was very lucky. Actually, none of my parents had high school, called gymnasium then. You didn't need even that to become a pharmacist at that time. My father had what in Norwegian is called middle school, which I guess is ten years. And then he had to spend some time training for a trade, like for a kind of skill. And after that he had to have about half a year, or something of that order, of university. Both my mother and father were very fond of reading. And I remember they bought books on auctions in Copenhagen. We got a big box with books, which could take maybe a few hundred books. You bought them sort of unseen. So I learned to read before I started school, and I learned to read in Danish. We read all the time, and we read a lot of garbage. My mother and my father read good books as well.

We were better off than most people. There was no question that I was going to go to high school. There was no question that I was going to go to university if I wanted to. They expected us to do that.

6.2.2 Applying for the University

I went to gymnasium at Hamar where I had a good teacher both in physics and mathematics. I always knew I liked technology. But it was a very unusual time since it was at the end of the war. So the teachers were arrested and the school was closed down. And at that time, when you finished gymnasium you had to have the best grades, otherwise you couldn't get in any places. I applied to The Norwegian

Institute of Technology (NTH) in Trondheim to become an electronics engineer. But my grades weren't good enough. I applied to become a chemical engineer and again my grades weren't good enough. When I say my grades were not good enough, I had a T ("satisfactory") in oral French and a T in written New-Norwegian. With those two T's you could not be an electrical engineer!

So machine engineering was the only thing I could get into. Actually, I probably made a mistake, because I also applied to go to the Danish engineering school in Copenhagen, and I got in there. And I was very torn about where I should go. But at that time I was brainwashed in a sense. Everybody in Norway at that time knew that the Norwegian Institute of Technology was the best university in the world. That was an established fact, even if it was so wrong it was unbelievable. But we all believed that. So I went there, and basically I was not interested.

I had to work for a year before I could be accepted. You had to have twelve months of practical training. Basically that was good. That year I learned an awful lot, as a toolmaker in a foundry. This was relevant to machine engineering.

When I worked in industry I worked for Raufoss Ammunitions Factory. And I recognized right away, that engineering at that time was like a union card. If you were an engineer you would go to some place where they make things and you were not going to decide anything. You were going to be a manufacturing engineer, at least as a mechanical engineer. There were very few people in Norway, unfortunately, that actually designed things. I felt, and the people I hang with felt, that if you got your degree you'd run your little factory somewhere. And what you needed was common sense, but not high mathematics.

6.2.3 Interests

Due to my lack of interest in machine engineering I played a lot of chess. I was a frequent guest at the Student Society, Studentersamfundet, and would be playing chess and became the chess champion. We played billiard and I became the billiard champion. We played bridge and I became the Trondheim champion in bridge. So I had a lot of other interests, including girls. And then we played a lot of poker, but we were just changing money. No championship in poker.

In my studies I had big problems, even though I finished on time. In the second year we had an important math exam. At that particular time the examination in mathematics lasted for a good six hours. That was after two years. You had one exam in mathematics, that was it. First of all I hadn't done anything, so I said, "I'm not going to do this." I hadn't been in the army, so I would go in the army and come back and do it next year. Because I knew I had to work. So a friend talked me into doing that later. He said, "we are going to have an examination experience. Just do it for the experience." So I did that, I went for it and I got a 4.0 in both mathematics and physics. That's as you know, the passing grade. And all my friends flunked in one subject or another. I was the only one to pass them all.

The good part of that story is that when I came here to Schenectady to get a job, the guy looked at my records from Norway. Of course he couldn't read it, but he could read physics and mathematics. He said: "In math you're a 4.0 and in physics you're a 4.0, that's very good." In the United States 4.0 is the best you can get. That's a perfect score. Then he looked at me and said, "In mechanics you only got a 3.5, you didn't do so well there." So there you are. I consider myself an honest person, but this time I didn't say anything. That's a true story.

6.2.4 Lodging and Job

After my studies, the reason it went as it went was as follows: After I was finished I was in the army in Norway for about a year. My wife and I were married at that time, we had a son, but we could not get an apartment. It was just impossible in postwar Norway.

I got a job in the Norwegian Industrial Property Office, dealing with patent matters, but we could not get a house anyplace to live in 1953. We wanted to live in Oslo, so I lived with a friend of mine, and I applied in the main newspaper, Aftenposten, every day. I always wrote I had no trouble with money. I didn't get a single answer. No answer. And I lived with my friend who actually lives in the United States now. And he said, "Go down to the Norwegian housing authority and register yourself." So I said, "No that's crazy, the waiting list is seven years. I'm not going to wait here for seven years." He said, "Go down and register anyway." So we went down and I registered. And then the guy looked at me and said: "Oh, you're married, where is your wife?" I said, "She's up in the country with her parents." He said, "You can't register, you don't live here." Then I recognized that I had lost. And that very day I decided, this is it.

We decided to go to Canada first. The reason for that was first of all that my wife had a sister there, and the second reason was that you could get a visa within three weeks. For the United States it took like a year to get a visa. I had no goal to go to the US. I just wanted to go, so we left. It took about three weeks to get a visa. It was great. But after I started working for Canadian General Electric, I recognized that in the United States, which was just across Lake Ontario, people got 30 % more than they did in Canada. And there was no reason why I should live one place or the other. But Canada was a really nice place, I really liked it there. There was no problem with housing. We had started out in Toronto, but when I got the job at Canadian General Electric it was up at Peterborough, which was a relatively small city with about 40 000 people at that time. General Electric was big, they dominated that city. But it was a great place.

However, being inside General Electric didn't mean it was easier to move to the United States, because this was during the Korean war, and everybody knew that money was much better in the United States. So the Canadian General Electric didn't want to transfer anybody. So they said, "No, no." And if you went by yourself, people were afraid they would be drafted into the Korean war. Nobody wanted to do that. So General Electric did not give transfers. Still, I quit, even though I had not been promised a job here. They said, "We can't give you a job, because it would be illegal for us to do that. But we're probably going to hire you if you come down." So I wasn't too worried. But I knew our personnel manager in Canada very well. And he came to see us for the train lift. And as the train left, he said, "Ivar, you're transferred!" That was a great relief, you can imagine, with a wife and a child going down here to Schenectady and not knowing anything would be hard.

I was accepted on a General Electric course for engineers, since they thought that the bachelor's degree was not good enough, and so they selected some people who wanted to do the extra work. We got lectures four hours a week, and we had to do some big problems on our spare time. And so we did. In Canada they had an A-course, and when I got into that I recognized that I hadn't done anything when I was a student in Trondheim. And so this was my last chance, I got a second chance. You asked me before if I was a self-made man. We worked exceedingly hard, my wife and I, for the first year in Canada. We got down here to Schenectady and we took on the next two years. I would say that I probably spent 40 to 50 hours extra on reading and studying and working. Not that I felt bad, I felt bad for my wife and my child. But the fact is, when I got into it I loved it, the stuff I hadn't done.

Also because when I got here for the B-course, I met two people in my class who were in a school called Cooper's Union, a school in New York City who takes the best people who apply, and the school is free. Absolutely for free. First of all, in New York City, there are a lot of people, and they can go for free to an engineering college. They get a huge number of applicants. And when I met those two people coming down here I couldn't believe it. I couldn't believe that someone my age could know so much. This to me was an awakening. It was a challenge. If they can know it, maybe I could do it too, if I worked hard enough. I worked very hard. GE gave this course, and the idea with this course was a very good thing.

When I came to Schenectady I worked for a man called Poritsky, he was a PhD mathematician. He was working for GE and did all kinds of math problems for people in the company, and I was his assistant. Then I went to the next person who was a German mathematician who had the same kind of job in steam turbine. These people took the young engineers, and if they could, hack it or be helpful. So you worked like four months with each. I had four or five of these assignments.

So, I got educated by these people internally. Plus I took this course. So I sought these people out when I came down here. I was warned of Poritsky. They said, "Oh no, he's so difficult to work for." I felt, "that's why I'm here," I wanted to.

And so, that was very helpful to me. I had no idea where I was going to go. I will explain the reason I went into physics. I had one assignment at the research lab, and I saw what people did there. That looked very intriguing to me. Because they sat in a window sill and discussed funny problems. It was one of my biggest discoveries. That you could be paid for doing physics. When I grew up in Norway, I had no idea.

When I grew up, going to the University in Oslo meant you were going to be a high school teacher. I didn't want that. So I discovered here that you can actually get paid for doing physics. Later, I worked with Hans Bueckner, who was a very famous mathematician, really amazing. He actually went back to Germany and came back here. He was really a brilliant person. And we worked in parallel with a problem. This was common. You didn't have a computer, you had to write things down on a piece of paper. Square root of two and things you missed, you got to compare. And I worked with him for maybe four months, and every once a week we compared our sheets. And when he didn't agree he calmly took my paper and said: Let me see where you are wrong. I swore I shouldn't be wrong. I was always wrong, he was always right. So I decided I would like to do something easier than applied mathematics, so I went into physics instead.

6.2.5 Quantum Mechanics

I was an engineer first of all. I was succeeding and lucky. I got the job at the research lab, and I felt very guilty up there because you didn't have to do anything "useful." I mean, there were nobody telling you what you should and shouldn't do. I was working with a man, my good friend, John Fisher, and he was my mentor at the lab and he was very good at doing that.

John Fisher, when I came to the lab, said what he wanted was to do quantum mechanical tunnelling. He probably knew about Leo Esaki's work where he had done tunnelling in semiconductors. I don't know if he did or not.[2] And after I had been there like a few months he said we were going to try to cross thin films. He said we have thin enough layer to go through, so he suggested that problem to me. I worked on this, trying to do it experimentally. And it's true, when we did this I had never had quantum mechanics. I did not believe that tunnelling was possible. So I said to myself, "You can't argue with your mentor, so I'm going to show this guy that this is not true." How can it be, when you look from a classical theory, with the electron bearing negative energy? That doesn't make sense.

At the same time, at Rennsselear Polytechnic Institute (RPI) I took quantum mechanics and I learned about tunnelling, and I still sort of didn't believe it. But then I did my experiment and I actually found out that it was true. And then I believed it.

Now I gave a talk at the lab. It was my first talk and I was very nervous. I remember that. And there were maybe fifty people in the audience, all distinguished scientists. And here I was, I didn't have a PhD degree and had done some work on tunnelling, convinced that I was right. And so I gave the talk and they were polite and they asked questions. "How do you know this is tunnelling, why isn't it ionic conduction, why don't you have a hole in the barrier? Why isn't it something else?" And of course I couldn't answer that. So I then spent the next few months thinking about how I could absolutely conclusively prove that it was tunnelling. That was my idea. While I did that I also took courses at RPI, where I am now. And my professor, whose name was Bill Huntington, talked about superconductivity in a solid state course. And he told us the resistance was zero. I did not believe that, no way it could be zero, maybe small, but not zero.

Now I actually believe it is zero, because I understand. But at that time I didn't know about the supercurrent and the quantum of flux. So I didn't believe it. But then

---

[2]He did not, according to information John Fisher gave the author.

he sort of mentioned in the class that there was a new theory by Bardeen, Cooper and Schrieffer, and this theory said that there is an energy gap in the superconductor. Right away, when he said that, I recognized that is what I need to show, that I have tunnelling. Because of the gap. So I went back up to the lab the same day, and I asked three people, Charlie Bean, Walter Harrison and John Fisher about this energy gap, how big was it. I had no idea, it could be a "mile high," a microvolt or whatever. It turned out to be a few millivolts which is right in the way with the tunnelling I had done. We talked about it, and they all said that it was probably not a good idea because it wouldn't show up, but they all said go ahead and try it.

Allow me to sidestep a bit here: Charlie Bean was my best friend. Unfortunately he died several years ago. But Charlie and I had lot of interactions. He was a great person. He was a remarkable person. Charlie, forgive me. But to be fair, Charlie knew all about magnetism, and when they started MRI, he said that they will not see fine enough details. You can't get down to more than centimetres, or so, he believed at the time. And so GE did not get into that. And Charlie was an expert in this. And then Paul Lauterburg published the image of an orange or whatever it was, where you could see the seeds. And that changed everything. But Charlie was one of the first people who worked on making high field superconducting coils, up to ten tesla. Charlie has always been a great help to me.

6.2.6 Tunnelling

I chose aluminium which has a nice oxide on it, and then on the other side of the barrier I had lead, because lead is superconducting at seven degrees. When you cool down with liquid helium you get 4.2 degrees. So that should really be all you had to do. The good thing being at a place like the research lab at that time, is that there were experts on everything. And I had never ever done superconducting work. After I got a "go-ahead and try it," it took less than a week before I had done it. Even though I had never done superconductors before.

And there are things you can do that you don't recognize. The sample was made of glass with aluminium and lead, and I put it down in the liquid helium with copper wires. And people said, "you can't do that," and I said, "why not?" They used manganese wires with low thermal conductivity. After I had done it with copper wires, everybody else started using copper wires too. Because it didn't make any difference.

In the first experiment I did, the aluminium oxide was too thick, so that didn't work. So I made one very fast, and then it worked perfectly. I had clear expectations. I looked for the energy gap. And as soon as lead superconducted, I found it. So there was the energy gap. And of course, being up there people said, "Oh, now you can put on a magnetic field and the gap will disappear. You can cool down more and put a pump on the helium and get it bigger." So it was one of these exceedingly lucky things, it really happened because I happened to be at a great place at that time, General Electric. Unfortunately, now they have changed. I felt myself very lucky.

People were excited, and I certainly was. Roland Schmitt was my manager at the laboratory. He wrote an article in Physics Today [1] that got quite a bit of attention. It was a nice thing. I studied now at RPI and I had to stop. As my friend Charlie Bean said, "You can't go to college, as it interferes with your education!" So I had to postpone it. And then later I wasn't going to go back, but my wife forced me to go back. She said, "You've spent so many years now, you have to finish." So I'm glad I finished.

After that came the measurements of the density of states. That came easy. The structure of the energy gap was known from the Bardeen-Cooper-Schrieffer theory. So I just measured that. But people said that tunnelling wouldn't be proportional to the density of states. The facts are that in the superconducting state, tunnelling is directly proportional to the density of states. If you tunnel into an ordinary material, tunnelling is not proportional to the density of states. And it has to do with effective mass and all sorts of things. But I didn't know too much about that, so I only knew that if tunnelling was proportional to the density of states, that's what you should see. And that's what I saw. So to me there was clear experimental evidence.

What further happened was, after I did this, I worked with a man called Karl Megerle, and he had a bachelor's degree in physics. He was a great help because he had good hands. So Roland Schmitt brought him up and said I should work with somebody, then you get more done. Then I worked with Howard Hart, and he's a great physicist. He was a great help. We did a paper where we did tunnelling down at 0.2 degrees or so, using He3 refrigerator. So I got a lot of help.

6.2.7 Reaction on the Paper

After we published the paper, the first person who invited me is a good friend of mine now, Elias Burstein at the University of Pennsylvania. And I went there to give a talk. It was the first talk I ever gave at a university. And when I got there it was a huge snowstorm. So I couldn't even get to the university. Finally, in the afternoon I gave the talk to Eli and two or three students. Even the subway in Philadelphia was closed that day. Later I gave a lot of talks and received lots of invitations.

My later skills as a speaker, which people sometimes give me credit for in terms of being somehow easy to follow, I owe first of all to John Fisher, my mentor at the lab. He is very good at that. And secondly, and this is really crucial, when you go to a lecture, very often you don't understand what people say. So the person who talks, doesn't do the right thing. I feel that I am a relatively smart person. If I can't understand what he is saying, it must be something wrong with him. I would say, now being a professor, I try to make sure that people understand what I say. But very often, I would bet you that in a class 90 % of the students have no idea what you are talking about. And that is something we should try to do something about.

I believe that saying everything and not missing anything in a talk is not enough. Analogies are a good way to work. Do something that people are familiar with. It is very difficult, you know. I know personally when I go to lectures, a lot of things

I can't understand because the person doesn't give me the right background. I try to let people know what I am talking about. I'm a more practical person from my engineering background.

Now, in spite of this I chose a theoretical subject for my thesis. This was because I at that particular time was very fortunate to be at RPI, because RPI and Syracuse were the only two schools in the United States who allowed you to do a PhD without being a resident. So I happened to be here; so that was one thing. Secondly, since I worked at GE, if I had to do an experiment, I had to do it at RPI. But I didn't have any equipment over there, so it would be easier to do a theoretical thesis. This way I was not tied down to any place. So that sort of came naturally.

Basically, when I did my measurements, Tinkham was another person who had studied the energy gap, using infrared spectroscopy, and he had some answers. Charlie Bean told me that, and I was a little nervous because my answer didn't agree with him. And Charlie said, "Don't think about it, now Tinkham must agree with you! Don't worry about agreeing with him, because this is the way it's supposed to be done." Infrared, of course, is a very difficult experiment to do. And this was easy. I did a lot of things in superconductivity. I did the tunnelling, and later on I confirmed the AC Josephson effect. That's when I got my Buckley prize. Then I was the first to do what I thought would be a big thing, the DC- transformer. With the flux flow, I measured that directly. I got a great kick out of that, because I sort of figured that one out. So when you've figured out how to do it, you actually make the samples, and... my god it works! I didn't believe in flux flow either, that's why I did it. If it's true, this and this should happen.

The citation for the Nobel Prize emphasizes the energy gap measurements. Basically I got the Nobel Prize because I confirmed the BCS-theory. So that was it. And I think that Bardeen was a great help in that case. This is how they work, sometimes the committee asks other people and it depends on who suggest you. Bardeen of course, had a lot of power, so people listened to him. And so he liked this experiment very much. I think that was quite helpful.

I was embarrassed in a way. Anybody, in principal, could have done this. You're just lucky and you have one good idea, and it happened to work. And that's what it is. When I give talks, I'm trying to tell people that is what they should look for, it can happen to anybody. I've had the fortune or misfortune, whatever you call it, to be a Nobel Prize winner. And some people have been desperate for it all their lives to get a Nobel Prize, and they do an enormous amount of work. And they finally succeed, particularly in biology. They are very clever people, but they are very ambitious, and you know... It's not that they don't deserve it, but also people do a lot of work, and they don't get it. And I was just standing there and it lands in my lap, you know, without doing anything. So it was really lucky. Life isn't fair, and I'm very happy that life isn't fair. Others say to me that it is fair because it was there for everybody, and I took the opportunity.

But anyway, it really clinched the BCS-theory, and at that particular time people like Felix Bloch, which I got to know pretty well, did not believe the BCS-theory. Even after superconducting tunnelling he didn't believe the BCS-theory. He had his own theories. Every theorist had his own theory, and didn't like it to be.

6.2.8 The Josephson Effect: Seeing, but not Recognizing

I wrote in my Nobel Prize talk that we observed the effect which later turned out to be what is now called the Josephson effect. This is true. Karl Megerle and I started making tunnel junctions out of lead oxide and tin oxide, which is thinner than aluminium oxide. And then very often you see the Josephson effect. And every time we got a sample that showed a so-called short, we thought it was a hole in the oxide, so we threw it away. I can remember distinctly that we had a big magnet to look at the energy gap as a function of magnetic field. With the big magnet turned all the way down there was really no magnetic field. But as soon as we put the sample in the field the "short" went away. And we say, "Oh, it depends on magnetic field, but how can this happen?" It happened anyway. And I don't say this because I want to take credit for the Josephson effect. I say this because, to discover something, you have to understand what you do. You can't just measure something looking like a short, throw it away and later come back and say it was the Josephson effect. You have to recognize what you do. If you don't recognize it, then of course you haven't discovered it.

But we were close. But actually, one of the curves we observed we published, with the "short" in it. But again I'm saying this as a lesson to people, that you have to understand what you do. But then I have the patent on the Josephson effect. I have the patent thanks to GE, on separating two superconductors, one Ångstrøm to a thousand Ångstrøm. So you couldn't make the Josephson effect without that patent. But we didn't know about that. And Brian Josephson was a smart person.

6.2.9 After Superconductivity: Biophysics, and Some More...

Author's note: Ivar Giaever changed his field of research to biophysics in the late sixties. We asked whether he would regard biophysics as even more complex than his original field, superconductivity. His response:

It's different. First of all there are no laws in biology. All the laws are physics and chemistry laws. So biology is mechanisms. There are a large number of unknown things in biology, because it's so complex, therefore very difficult to do. It's a most satisfying thing that you can find out new things. Physics, I think, has played its course.

I was just in Israel in the year 2000, I gave a public talk, but it was part of a meeting on string theory. String theory is something that has gotten to me, people now have worked with string theory for roughly 25 years, and there is no comparison with experiments. Absolutely none. That upsets me, because physics to me is an experimental science, and if you start discussing how many angels can stand on the head of a pin, there is no limit to that. It's amazing that all these people can do all this theory but with no connection to reality. That to me shows that there is something wrong with physics. And the way I feel about physics I tell to students at RPI is that we are moving into a new paradigm. Physics now is just about using what we know in different ways. Like you can invent a laser. You can invent Josephson's effect. Inventions are going to be the important things. There are not going to be any new laws. To speak of anyway.

If you think about electromagnetic theory, Maxwell's equations were written back in 1878, and they are still there. There is no way you can get a law associated with your name. But in many ways you can make an invention. Now people look for different kinds of particles, which we believe is part of dark matter. But these particles, some of them don't interact with ordinary particles. So people are actually looking for particles that don't interact. It's fine to do that, I mean if you find them, of course you become a hero.

But the fact is, a few years ago they dealt with the tachyon, a particle going faster then the speed of light. I mean you can do these things but I don't know if somebody should pay for it. It seems to me you can be too wild. You can be too far out.

Ed Witten gave a talk, and I was part of a small audience. He gave a good talk actually. Among other things he said, which I liked a lot, "There are three ways of doing string theory." He paused, and then looks up, and says: "I wonder where the other two universes are?" And then it turned out that you could combine it into one way. Then he was finished and we could ask questions. He said that if the theory was right, then Einstein's theory sort of wouldn't be right. So I couldn't help but raising my hand, and said couldn't we do some experiments? And he looked at me like if I was an idiot, I think it could take more energy than the Big Bang to do this kind of experiment. It's just outside of what is possible. And I don't like that part.

6.2.10 Vision?

Basically what we're doing now, we have this method on how we can look at cells in different ways. And we try to apply this to different things in biology. That's where the physicists can be very useful, because we have all these different ways we can invent to measure things. And biologists don't have any ideas about that. They come from a different experience, so they will buy equipment, but very rarely do they make their own things to do something. Physicists have learned that if there are no equipment to measure this, you have to make some equipment to do it. So I think physicists have a lot to give to biology. What they have to do is to work with biologists to learn what their problems are. You see, biologists' workdays are full of things they don't understand, and the physicist's workday is completely the opposite. He understands everything, he's worrying about what experiment can he do now.

Biologists don't have that problem, physicists have the problem that there are no really important problems for them anymore. And when people discovered high temperature superconductors, the whole physical society was very much interested in high temperature superconductors. Because whatever they did, it occupied them.

I'm very happy I've made the switch. I've never regretted that. Actually, it's good to change, that is another thing John Fisher told me. It is good to change your field, because if you work in a certain field, like when I worked in tunnelling, all the experiments people do, I had actually thought of. I didn't do them all, because I didn't have the time, or I didn't think they were important, but then they turned out to be important.

Then you get to the stage were you don't want to think about it anymore, because you have thought about it. When somebody asks you a question you say: "No, this is the way," because you have already thought about it. But then suddenly if you learn something new, if you change your field, like when I went into biology, there were lots of things I hadn't thought about. Lots of interesting new things to learn. It is very difficult if you have learned something to go back and look at it with new eyes, because you have made the pattern in your head and you know how it works. And therefore you're stuck, and therefore very often somebody else discovers something different. They come from a different background, and they see something that there is no way you can see by yourself.

It is possible in biology or biophysics to start, knowing very little, and learn step by step. I happened to go into immunology first. It was a wonderful thing to do because immunology is so much like chemistry. And I invented a method by which you could see, not single molecules, but single layers of molecules with your naked eye. I was very disappointed that it never became a practical method. You can invent things, but you have to have a problem. When you have a problem, you have to work on the problem. You don't have to know a lot of things, it's just not true. If you know too much you won't do the problem. But a lot of people have said to me, "How can you put two metals together with an oxide layer between them, separating them. We thought it would be a short or something, you can't do that, it can't be done." And personally I didn't believe it when people came up with the tunnelling microscope, because I tried that myself. I said, "that is wrong." Well I was wrong in that case, because I had thought about it, tried it and didn't think it could be done. And so I didn't believe it when they showed it. So I was wrong, and it's ok to be wrong. But you see, there was no way I was going to make the tunnelling microscope, because I had decided it's not going to work.

Reference

1. R.W. Schmitt, Phys. Today **14**(12), 38 (1961)

Chapter 7
Brian D. Josephson: Cooper Pair Tunnelling: The Josephson Effects

The case of the student against the Nobel laureate.

Fig. 7.1 Brian D. Josephson with the Physics Today cartoon on the controversy with Bardeen

7.1 An Abbreviated Account Based on an Interview and Available Literature

Brian D. Josephson shared the Nobel Prize in Physics for 1973 with Ivar Giaever and Leo Esaki (see Chap. 6) *"for his theoretical predictions of the properties of a supercurrent through a tunnel barrier, in particular those phenomena which are generally known as the Josephson effects."*

Josephson grew up in Cardiff where he was born in 1940. He was mathematically inclined at a very young age. He started studying algebra by himself. His father did some mathematics and had some mathematics books. He followed them, and worked much on his own. He entered the sixth form, grammar school, at age 11 in 1951. Mathematics was the kind of science that interested him, and he had a physics teacher who helped him with theoretical physics textbooks at university level. So, he did advanced level physics and even chemistry after one year in the sixth form, going ahead by himself. He took mathematics at A-level two years in advance, and physics and chemistry one year in advance.

Josephson started studying at Cambridge at the age of 17, one year earlier than normal. He could have gotten in even earlier, having passed the exam a year before. But he was told it would be better if he did not come at such a young age. The policy was different at Oxford, where some people were allowed to enter as early as the age of 12. He started off with mathematics, and did a lot of mathematics for two years. Already here he wrote a paper which he submitted during his second student year, a theorem that was different from what he found in his textbook.

He then changed to physics in his third year. During the course of that year, there was an experiment going on at Harwell, measuring the gravitational red-shift using the Mössbauer effect. At the age of 20, after hearing a lecture by Robert Fish about the Mössbauer effect and the experiments, he tried to reconstruct for himself a theory about how the Mössbauer effect worked. He tried to explain the Mössbauer effect by considering what happens if one adds a nucleus in a potential well. He found that there would be a shift which was dependent on the temperature. His letter to Fish on the matter was passed on to Ziman who sent it to Walther Marshall, who contacted young Josephson and arranged for him to go to Harwell and write a paper. This became his first publication in physics, in his last year as an undergraduate student at Cambridge.

Josephson became a graduate student at Cambridge in 1960. In his official PhD-project he was doing some work on experimental superconductivity under Pippard, who suggested that he should learn about the theory of superconductivity. So he got to understand the BCS-theory, including both Anderson's version, and Gorkov's, on how superconductivity works. The paper by Anderson on the pseudo-spin model for superconductivity was important. He started to think about the analogy with magnetism, and that spatial direction was like a phase, and thought that it would have a real existence. He convinced himself that the phase of the wave function was important. That was present in most of his approaches.

Josephson worked with group theoretical considerations to conceive conditions under which phase would matter. He credits Phil Anderson with the insight he had gained in the concept of broken symmetry, which turned out to be important, and already present in Anderson's 1958 pseudospin formulation of superconductivity theory. Anderson was present in the group and giving a course on concepts in solids, resulting in a famous book on that subject. In the BCS-theory it had been shown that in a superconductor there is a strong positive correlation between two electron states with equal and opposite momentum and spin. Anderson showed that in the case of perfect correlation the system can be described by a set of interacting "pseudospins." Each pair of electron states may be represented by a pseudospin, with its spin in the positive z direction when both states are unoccupied, and in the negative direction for both states occupied. Other spin directions represented superpositions of these two. In the case of attractive interaction between pseudospins there exists a state of lower energy in which the pseudospins are tilted out of the z-direction with increasing amount as one goes from states below the Fermi surface to states

7.1 An Abbreviated Account Based on an Interview and Available Literature

above it. The pseudospins can lie in any plane containing the z-axis, all representing the same energy. This situation is said to break the symmetry of the original state, and causes superconductivity to occur. The angle of the plane in question becomes important and relates to the phase angle introduced by Gorkov earlier. At one point Josephson found it could only be the *phase difference*, and not the absolute phase that would matter. But the phase could only matter if there was a transfer of Cooper pairs through the contact.

In discussions with Brian Pippard, the question of the effects of the coherence factors in the BCS-theory was brought up. These aspects motivated him to do the calculations properly, taking into account everything which was in a paper by Cohen, Falicov and Philips, who had a way of calculating the current. So he basically applied their method to the two-superconductor case. This gave the phase difference dependent factor, which he was expecting, but also the zero voltage current, which he was not expecting.

Josephson says he has a rather vague memory about which events followed what. He did not have discussions with Phil Anderson about the actual formulation of mathematical details, because it was based on other works. He believes he may have had discussions about the broken symmetry concept with him, but without input into the calculation. As far as Anderson's influence is concerned, the broken symmetry in his course was crucial, both his published paper in Physical Review with the pseudospin, and in his lecture course. But on the other hand, the idea of the phase as something that would occur in a tunnelling experiment, that was definitely his own.

His original paper was published before writing his fellowship thesis. The analogy with rotational objects, as included in his Nobel lecture, helped him understand, but he got rid of it in the actual published paper. There were other significant results he published, like the coupling free energy, and later he wrote down the basic set of equations for the junction.

Pippard did not have much faith in the phase as an important aspect. He was quite sceptical. So Josephson felt it was good that Anderson was there, because he encouraged him and said "That seems to be right." Josephson confirms Anderson's claim that he had to defend him in teatime discussions. One thing that seems to have been relevant at the beginning, was the flux-quantization experiments by Deaver and Fairbanks around 1960, where they used a suspended fibre covered with superconductor to measure the magnetic moment. That showed that the phases were a real thing, and were continuous when you go around.

One thing Pippard was interested in, that might have stimulated Josephson according to his recollection, was the junction between a superconductor and a normal metal without a barrier between. How current got from one to the other. Atkins and Josephson tried to confirm the effect experimentally. But their efforts turned out to be unsuitable both for dc and ac cases.

When Josephson's paper came out, Bardeen was negative. They met the first time at Queen Mary College in London at a low-temperature conference. Before that, Bardeen had added a footnote in proof in a paper of his, referring to Josephson's paper and saying it was wrong. So a special session at the Low-Temperature Physics conference at Queen Mary College was set up, for Bardeen and Josephson to present their opinions. Bardeen started off first. Josephson interrupted him at some point and objected. Bardeen continued. Then Josephson gave his own version, and was supported by de Gennes who asked about locality, where de Gennes had done some work. So Bardeen had this—in Josephson's mind—fallacious argument that pairing would not continue through the barrier, causing decoherence. Josephson's view was that just exponential decay wouldn't change the phase, so it would not decohere. Bardeen then, when the experiments were done, realised that his argument must be wrong. So he produced a modified version, where he said he forgot a term, and got a different result. But then, by the time of the Colgate conference in 1963, he had given up his approach altogether, and made a speech saying Josephson was right and he was wrong.

The situation with "the student against the Nobel laureate" became a famous story. It entered the pages of Physics Today with a nice cartoon. It could have caused some distress in the student, but Josephson says it didn't because there were people on his side. Rather, the main concern was whether there might be some fact that would suppress the effect. In fact, the experiments in Cambridge did not yield any supercurrent. Anderson worked out what the problem was. It was caused by the electron leads which gave electronic noise. Josephson had worked out that the thermal effects would be important at low temperatures. The successful experiment was done by Anderson and Rowell at Bell Labs. On a historical point, Josephson points out, Anderson's calculations for the magnetic field dependence was wrong. Josephson was not sure if this is on record, but thinks Anderson corrected it in proof after Josephson sent him a letter.[1] The point was, according to Josephson, that Anderson realized that the flux was the important thing, but he had only taken into account the flux in the barrier, and not the flux in the penetration region in the superconductor.

According to Josephson's recollection, his first paper did not explore the field dependence very carefully. Fisk got the curves that showed the oscillations in the field. The time when the experiment by Anderson and Rowell came out was about nine months after Josephson's paper, which was published in June '62.

Regarding other work, in a more trivial sense, there was Ambegaokar who discovered the extra factor 4 in the calculation. He thought this meant Josephson's method was wrong, because he did it with Green's functions, and meant Josephson had left out this factor. But Josephson's method actually gave the same answer as his. Anderson may have calculated the temperature-dependence. Josephson did his work at absolute zero. Anderson may have done the temperature dependence for the equal gap case. Josephson derived the general field equations, the spatial dependence, and also the effective capacitance. So he got the oscillation, the plasma resonance, as he called it.

Afterwards Josephson went to Urbana, Illinois, after getting his PhD, and there he got into critical phenomena with Kadanoff. He got involved with two things. One was the question of the connection between the superfluid density and order parameter. And then there were also some inequalities for the specific heat.

---

[1] Anderson acknowledges this story, see the Anderson chapter (Chap. 8) of this book.

7.1 An Abbreviated Account Based on an Interview and Available Literature

Josephson was also involved in some unsuccessful experiments, and has some ideas why they weren't successful. He worked with John Atkins on tunnelling. Atkins was involved because he had been making junctions, so he was used to that technology. According to theory the Earth's magnetic field was about the size that would have significant effects. So to get as low fields as possible, they had Helmholtz magnets that would compensate the Earth's field. They could detect nanoamps and critical currents. And there was no non-zero voltage current down to nanoamps, or maybe less. The reason turned out to be the thermal effects. They needed a lower resistance and they used aluminium to reduce the risk of short-circuits. The approach at Bell was that they used lead with junctions that had low resistance, so they showed up. Atkins and Josephson used aluminium, and didn't find it. And as far as the ac-effect, they tried to couple two junctions together, one was source and one was detector. Why that didn't work still isn't too clear to Josephson, but the amount of effect must just have been too small to be seen. He believes Ivar Giaever may have used a configuration where there was more coupling. It was more ambiguous, but it was the way to do it. "Better to do an imperfect experiment that works, than a perfect experiment that doesn't" is Josephson's comment.

The results were eventually published in J. Phys. F.

Chapter 8
Philip W. Anderson: Superconductivity from a Broader Perspective

"I ran into superconductivity through being very friendly with Bernd Matthias."

Fig. 8.1 Philip W. Anderson

8.1 Biographical Notes

Philip Warren Anderson shared the Nobel Prize in Physics for 1977 with Sir Nevill Francis Mott and John Hasbrouck van Vleck *"for their fundamental theoretical investigations of the electronic structure of magnetic and disordered systems."* Anderson's influence on condensed matter physics has been of profound importance. He is often characterized as one of the most influential minds in all of theoretical physics in the second half of the 20th century.

Anderson's initial interest in superconductivity came from association with the experimentalist Bernd Matthias at Bell Labs, with whom he first worked on ferroelectricity. After the BCS-paper came out he made a study of gauge invariance which they had not considered, and which was a concern among theorists. Here, he discovered what is today known as the Higgs mechanism (see below, and Chap. 12). Also, he was a key person in the development of the pseudospin formalism for superconductivity towards the end of the 1950s. This line of thinking has later been successful in completely different fields of physics. He contributed to the development of a theory for d-wave and p-wave superfluid phases of helium-3. With Kim he did highly original studies of the dynamics of quantized magnetic flux in superconductors in the early 1960s. He coined names like *dirty superconductor*, and *spin glass* and probably also the name *condensed matter*, and of course was the inventor of the theory for *Anderson localisation*. His stay in Cambridge 1961–62 was important in inspiring Brian D. Josephson to develop his theory for Cooper pair tunnelling between superconductors, the DC and the AC Josephson effects. From more recent years his efforts to create a theory for high-T_c cuprate superconductivity, the so-called RVB-theory, stand out as a major period in his career.

Phil Anderson was born in 1923 and grew up in an intellectually stimulating and outdoors loving college environment, with college teachers in the near family. After high school he had an intention of majoring in mathematics, but at Harvard things turned out differently. This was during the wartime, and electrical engineering and nuclear physics were important subjects. Anderson chose electronics and went to the Naval Research Laboratory in Washington DC to build radar antennas during 1940–43. Back at Harvard from 1945 to 1949 he enjoyed both the courses and the friendship of people like Tom Lehrer, the mathematician who turned into popular singer, known widely for his political humour. Anderson chose van Vleck as his thesis adviser due to greater accessibility than Schwinger. He married in 1947, and settled down to learn modern quantum field theory, which turned out to be useful even in experimental problems. This was at the birth of many-body physics, an area where he was later to be a major participant and leading scientist. Having completed his thesis in 1949, he was hired by Bell Labs to work with a number of outstanding scientists, like Bill Shockley, John Bardeen, Charles Kittel, Koyners Herring, Bernd Matthias, and Gregory Wannier. He ended his career at Bell Labs as Assistant Director 1974–76. At Bell, he also became acquainted with the work of Mott, and of Landau. Anderson became a professor of physics at Princeton University in 1975. Both he and his wife were quite active politically in the Democratic Party in the 1950s. They worked enthusiastically for the candidacy of Adlai Stephenson towards the presidential election in 1952, and were active in several other connections too.

8.2 His Own Story

8.2.1 Early Influences

My family, mother, father and uncle, had an academic background. My mother was the daughter of a math professor, my father was himself a professor of biology at Illinois, and my uncle was an English teacher in Wabash College. At the University of Illinois I knew a number of mathematicians and physicists and my parents were close friends with the head of the physics department, Francis Wheeler Loomis. He was famous really, I guess not any more, but he essentially started the department in the thirties, at that time totally in a backwater, and he built it up to be one of the greatest departments in the country. He hired Fermi students, people like Lee and

Goldberger. The Illinois tradition has always been that the department chairman is a dictator. They have usually had remarkable results, hiring the sharpest of people. The next thing he did, once he lost them, was to hire a group of solid state physicists, starting with Seitz and Bardeen. What he did for me was to gently suggest to my parents, since I did have a scholarship to Harvard, that it was important to take physics the first year. And that was the end of me. I disliked physics in high school because it was all qualitative, and I was just having fun with instruments. No explanations, and nobody really gave me any idea that people actually know how things work.

But at Harvard, during my first year in college, I suddenly began to understand how physics works, and I became very excited. And I stayed that way. I realized physics was my line, although I started out formally majoring in mathematics. But now the war intervened and we were all advised to take engineering physics, so we could be useful to the radar establishment; except for a certain number of "pet" people, of whom I was not one, who were quietly passed the word that it was a good idea to stay in nuclear physics. They went off to Los Alamos. And I went to a radar lab, the Naval Research Lab, at the southernmost tip of Washington DC.

8.2.2 Career Choice, Family and Politics

I guess the only urgent message I got from my family was, my father was only moderately successful as a professor. My mother always felt that he didn't get such a great salary. And she knew her father and her brother didn't get good salaries, so she said; why don't you go into industry? So I fixated on the Bell laboratories, which is both academia and industry at the same time. I didn't know that it was virtually academia, at the time, nor did she. By this time I was married; in those days a post doc fellowship didn't pay for a wife and a child. So unless I got a job... We were a little tired of living very close to the edge.

I was politically interested from the time of McCarthy-Truman actually, my first vote was for Truman—and later I guess especially the Vietnam war. We were very enthusiastic supporters of Adlai Stevenson of course, as Adlai Stevenson drew a lot of people, like me, into the Democratic Party. We were quite active then, as a matter of fact. We lived in this very backwater exurbanite community, very suburban, very republican. So it wasn't hard to become the Democratic committee for our tiny community.

My wife actually reached, I believe, the vice chairmanship of a state committee for the gubernatorial candidate in 1953, Bob Meyner. This was the year after Eisenhower was elected president. New Jersey elections are one year out of phase. She was actually vice chairman of something called Independents For Meyner, which was a fake, because we were perfectly solid democrats. And we were active in forming a democratic club locally, which had two distinctions. One was that a number of our members later became active in the state administration, and the other was that almost all of our speakers were later indicted for one corruption or another, because they were from the Democratic machine. There was a lot going on in those days, but I don't think that it was even roughly as bad as it was made out to be. But the republicans are very good at selling a minor peccadillo as a major one. And I'm sure that the democrats do the same thing to the republicans, every once in a while. My wife's co-chairman, I believe, was probably a member of the mafia. He was a very nice man, however. We also learned later that they were quite respectable as neighbours.

8.2.3 Electronics Physics and Harvard

Now to get back to physics, I was talked into electronics physics at Harvard. Pushed into it! I wasn't terribly interested. Some of the courses I enjoyed, and some of them were a terrible bore. One course was taught by an engineering professor who insisted on your notebooks being exactly correct, and that kind of thing. But some of it was fine, how electronics actually worked.

I liked radar, it had seemed a great mystery. It was fun to learn about it. When I arrived at Naval Research Lab, there was a six week course where you learned about radar, and I probably learned more electronics physics in those three or six weeks than I ever learned at Harvard. It was well motivated and very well organized.

My graduate school years were divided into two pieces. The first piece was absolutely free from the world. I had a stipend on which a single person could live reasonably well, enough for pocket money. I got to know a group of people who were not only physicists, but a very broad, diverse group of people from philosophy and English. Tom Lehrer was there as a mathematician, Chan Davis was a mathematician in those days, and a published science fiction writer. Bob Welker was an English major and a folk singer. Dave Robinson was a chemical physicist. It was just fun. I didn't do much on my thesis, till a year and a half. Well, that's the Harvard system; you take courses for a year and a half. There was the big quantum mechanics course. That was a marvellous course, and at the same time, before I ever had a formal course in quantum mechanics, Schwinger gave this entire course that was a kind of "core dump" about everything he knew about nuclear physics. Including Green's functions and nuclear moments. Schwinger was a marvellous lecturer, but not a very good teacher. One had to take notes as fast as one could and read them somehow between one lecture and the next. I learned an awful lot. We had a stiff curriculum. I had to do four courses at this kind of level, and get A or B in it. And that was not easy. I wasted some courses, or rather Harvard wasted some courses on me, but on the whole it was great. And I had a wonderful experimental physics course, with Oldenburg. He borrowed old equipment on which people had won Nobel Prizes and did famous experiments. He had an old oil drop experiment, and he used one of Bainbridge's old mass-spectrometers.

8.2 His Own Story

8.2.4 Encountering Superconductivity

I ran into superconductivity through being very friendly with Bernd Matthias. I arrived at Bell labs on the same day as Bernd Matthias and Gregory Wannier and a man named Jack Galt, not so very famous, but a very, very good physicist, better than many important ones. Bernd Matthias and I, since we were all the same generation, we got to know each other very well. That's the way it works at Bell, you come in and you socialize. Ted Geballe and I worked at Bell at the same time too. Bernd was actually not working with superconductivity, he was working on ferroelectricity, as I was, so we worked very closely together.

And then Bernd switched to superconductivity. In '51–'52 he worked in Chicago, collaborating with John Hulm. He came back, no longer interested much in ferroelectricity. Instead he went into superconductivity. I was not working with him, because I had become, by that time, kind of a house theorist for magnetic resonance and magnetism. And we had hired Harold Lewis, H. W. Lewis, who was a particle physicist, and a student of Oppenheimer, who then had been at the Institute of Advanced Studies at Princeton, because he refused to sign a loyalty oath at California, and we hired him to work theoretically with Bernd Matthias.

But that couple never really worked out very well, because Harold was very much a pure theorist, a very theoretical theorist, and Bernd was very much into empiricism. He infuriated Harold, and Harold infuriated Bernd.

I was used to Bernd, because I had some ideas about ferroelectricity, and I had written a couple of papers and he had some respect for me. I guess the first paper I co-authored about superconductivity was when I happened to be sitting in his laboratory because it was air-conditioned, and was chatting with him when he discovered his first alloy superconductor. So he put me on the paper. And I had nothing to do with it. But then Harold was the official superconductor guy. But I didn't think too much about Harold. Of course he gave talks and we all listened to what he had to say. He even did experiments of his own. He did a sophisticated experiment, which, however, didn't tell us too much about superconductivity.

8.2.5 Cooper

I heard Cooper when he was at Princeton before he went to Illinois, at least he was visiting the Institute. He came up to Bell Laboratories and gave us a little talk. He talked about his pairing idea. And we all said; yes, but there were problems with this kind of thing because you can seem to show for instance that you developed a bound state at the Fermi level; but then it turns out that statistics spoil the apparent singularity that you get. And that was the problem that bothered Walter Kohn, who had just been working on that. Kohn-Majumdar is a famous paper where they seem to show that the Fermi surface doesn't have any effect, which probably has had a more negative effect than any of Walter Kohn's other papers ever had. It's almost as strongly negative as his great paper on local density approximation is strongly positive. Because it discouraged people from finding all kinds of Fermi surface effects. It was right of course, but the further implications weren't. So in each case you had to prove that the Fermi surface actually had effects.

Cooper seemed to have something that really showed a Fermi surface effect. We all had in the back of our heads the Kohn-Majumdar paper, which showed that you couldn't have a Fermi surface effect. And so he went off to Illinois and gave it to John and Bob, and the rest is history.

And actually, it was David Pines who came to Princeton and gave a talk, very early on, about the BCS work. We saw the Phys Rev Letter, but David Pines came and gave a long talk. We talked with him some. At that time I really didn't know David very well, but had met him of course. And I rode down with Larry Walker and Harry Suhl and the three of us talked about the results in the car on the way home. And we developed ideas, some of them were Harry's, some of them were Larry's and some of them were mine. We developed this thing that looks a little bit like spins, somehow. So we began to develop what would later become the spin representation of superconductivity. I acknowledged Harry in the paper about it, but he said, "I didn't contribute that much." So he refused to be on the author list, which I offered him. There has never been any difficulty between the three of us about that.

8.2.6 Order Parameter

The concept of *order parameter* had been introduced by Landau. We had been working with a similar thing in antiferromagnetism three or four years earlier. Anyhow, then Wentzel who was a friend of Bernd's from his Chicago stay, came and spent some time during the summers as a consultant. And he also had heard of the BCS papers, and he said, "this can't be right, it's not gauge invariant." And he gave talks about non-gauge invariance. There was Wentzel, there were unpublished papers by Kohn, who had checked into it, and there were preprints on it. And so I took this idea that we had developed, from these discussions, and the fact that I felt that it would be gauge invariant, because it was obviously right.

The BCS team didn't worry about it. I wasn't really worried about it either, but I felt that it could be a big feather in my cap if I could actually do this. And besides, I thought it was interesting. I made one very long incomprehensible paper, based on these ideas plus some formulas borrowed basically from Dave Bohm, and some ideas borrowed from the Australian group, and I put together this very messy paper about gauge invariance. I think it did prove the point, but it certainly was not very comprehensible.

I wrote this alone, and I remember when I first got the basic idea, I was lying in the lovely fall sun in a field near the home we had at that time. I came down from that hill like Moses from the mountain, on fire from excitement. So I sent that off, and then almost immediately followed it up by seeing how to do it formally. That's actually when I inserted the equation of motion method, and the random phase approximation.

It attracted attention from BCS, and they were happy to have something they could add as a footnote to the big paper, and say that Phil Anderson has proved gauge invariance for us, so we don't need to worry about that. My first paper, the very first, was done in the fall of '57, and then this big paper was completed in the winter/spring of '58.

8.2.7 1959

When I was in Berkeley I heard some lectures by Leslie Orgel, and I realized how to do superexchange. I made a big paper on superexchange and the Mott phenomenon. It was finished in '58–'59. (This work later played a big role in the cuprate story.)

Since my earlier trip to Japan, the powers at Bell labs had decided that I have had this wonderful experience of overseas travel, so I didn't need to travel overseas anymore. This is just the way they were in those days. You didn't travel much, but when you did you went first class. But then, of course all my academic competitors were going back and forth over the Atlantic using MATS (military air transport-they had Navy contracts). But finally they liberated me, so I was allowed to go to Russia in '58, and to the Cambridge conference on superconductivity in 1959. Matthias got mad at me because he was not allowed to go. It was ridiculous that Matthias couldn't be allowed to go at that time. He was not allowed to go and I was supposed to represent him, and I was not about to, because I had lots of things to talk about. And I wasn't going to talk about his stuff, because that was materials science. Cambridge was the big international conference where BCS was legitimized, and accepted.

Then, on the way I had acquired a graduate student. David Pines had finally left Princeton because he was not given adequate tenure, and David Pines was hired from Princeton to Illinois. Pines had a graduate student from France called Pierre Morel. Pierre Morel was one of these mathematical wizards from École Normale Supérieure in Paris. He followed Philippe Nozières as David Pines' student.

Morel was the science attaché at the French embassy, or consulate, I guess, in New York, in the UN. He couldn't go with David when David went to Illinois. So David gave him to me. I had a couple of ideas for him. One idea was the fact that there could be angle dependent superconductivity that wasn't necessarily all BCS. But there could be p-wave and d-wave superconductivity and so on. And I had written a lot about that in my notebooks, and I set Pierre to work on that. At the same time I had these ideas of phonons, how to really formally express, not in an average messy way like the Bardeen-Pines way, but specifically, formally how to manage the phonon problem, and how to actually get a formal representation of the BCS-theory.

And I set Pierre on that, also. Eventually, one of the papers was called Anderson-Morel, and the other was called Morel-Anderson. I don't remember which was which. What I remember about Pierre was that already in 1959, he and his beautiful wife came to the airport when my wife and I set off for that meeting. We had taken a private plane from Morristown airport to Kennedy. They met us and our private plane, and presented us with a big batch of flowers, and then we went onto economy class on Icelandic Airlines, taking the lowest possible fare. So this was rather incongruous. But he was my graduate student, and he did wonderful things on two problems. The discovery he did was the anomalous superconductivity problem, which later became the He3 problem. He was the person who discovered that there are distinct phases, that the p-wave of one form is not the same as a p-wave of another form, and that a d-wave of one form isn't the same as a d-wave of another form. So we had this cubic shaped order parameter for He3, for d-wave in He3. And we also had an axial order parameter for the d-wave in He3. The two states have the same transition temperature, because that's a linear problem. The moment you get into the non-linear problem you have these different possibilities of different phases. So he made that fundamental discovery, that if you ever have an anomalous superconductor, it will be characterized by having different possibilities for phases—and that was the first observation to be made in 1972 on the real thing.

8.2.8 More He3, and Phonons

In '59 we didn't know that we were doing the theory of He3. It materialized later. We picked up He3 from Keith Brueckner, because Keith Brueckner worked at Los Alamos. He was a nuclear physicist. And he was aware, which we weren't, that He3 was the decay product of tritium. And because tritium decays in twelve years, it was just about the right amount of time after the hydrogen bomb to have macroscopic amounts of He3 available. So he knew that there were experiments on He3. He had been calculating what kind of superfluid phase might develop in He3. So he told us that, and we told him about the angle momentum dependent phases. He started publishing with his student, and I said, "no, that's not fair, I've got a student too." So we published it together. And then Morel and I wrote this paper about the phases, the different phases, about the real physical properties of the anisotropic BCS state. We called it an orbital ferromagnet, which is correct in a sense.

So this was 1960, and this was the first paper on superfluid He3. It was a busy time. And also in 1960, I went to a meeting in Utrecht and talked with Bob Schrieffer about my idea about phonons, and Bob told me about the Eliashberg theory. That what I was trying to do, in my usual way, picking up whatever I needed in terms of equations of motion, had really formally been worked out by Eliashberg. But of course, Eliashberg didn't know anything about the real phonons. He didn't know what kind of approximations needed to be made.

So Morel and I incorporated Eliashberg with the phonons, and we wrote the first paper on detailed calculations of the energy gap function, as a function of frequency. My basic idea had been to recognize that what was essential, was to average over momentum and keep frequency. And if you do that you have an interaction which is local in real space and long range in frequency space. And this is what makes energy gap spectroscopy possible. So that was Morel's second project.

The reason why you haven't seen him later is that he essentially became the space-tzar of France. He became a much more important person in the science hierarchy than Philippe. Nozières didn't envy him because Nozières preferred his own type of career. They always competed, and Morel got the girl they both were courting. Philippe was a better scientist. But Morel was no slouch.

Morel looked around and he realized that he had to compete with several bright people coming along. So he got out of condensed matter physics. It was too crowded. He's got great skills with people. I met him later, actually in Philippe's house last time I was in Grenoble.

8.2.9 Concepts in Solids

All of this happened before 1961, when I went to Cambridge on a sabbatical. I gave a lecture series there on *Concepts in Solids*, later printed as a book.

Brian Pippard was there, and his best student came to my course. That was Brian Josephson. He must have been about 20. He had been a child prodigy. He had a mother who drove him very hard. But he was also very bright. He already had one really important discovery in physics, as an undergraduate: The thermal effect in the Mössbauer effect, the thermal shift, a second order relativistic shift. All the students, including Josephson, were quiet. I couldn't get them to respond. He would only come up to me if I had been mistaken in the lectures. Or if I expressed something badly, he would come up and say "perhaps..."

You know, the Cambridge custom is that the graduate students socialize with the senior members of the group. Sit together at tea. Well, not exactly socialize, but you have that degree of cohesiveness. We had coffee and tea together. And as he was discovering the Josephson effect we were discussing it over coffee or at the tea table.

8.2.10 The Josephson Effect

Josephson came to me with an expression, very preliminary, a very messy expression for the tunnelling current between two superconductors, and it had this $\sin\varphi$ term in it. And he said, "can this possibly be right?" He would like to have somebody else look at it. I checked over the method. It had only one term, and it was obviously all that was necessary. All the other terms would cancel. I could see that it was right. I didn't really physically understand it, so Josephson and I and Pippard and a couple of other people sat around the tea table and discussed it. And finally I said, "this is $J_1 \sin\varphi$, it depends on phase and there is no reason why it shouldn't depend on the phase." The other thing was that I had said in my lectures that phase is a real physical variable in a superconductor, and it would be nice if we could measure it. I don't know if I used those words, but that was the content of it. And it is in principle measurable, it behaves like a physical order parameter. The macroscopic spin representation should mean that you have a physical order parameter. The relative direction in complex number space of that parameter is the phase.

The Josephson effect was published in Physics Letters, because he (or at least Pippard) was not quite sure it was right, partly, and partly because he didn't want to pay the page charges. After I left he wrote a thesis, a very thorough exposition of his ideas, and that was given to Trinity College (it was the application for a fellowship), and only four or five copies were made. And as far as Josephson was concerned, Trinity College was the world. So that was publication. But it was not publication as far as the rest of the world is concerned. Just because Trinity College has more Nobel Prize winners than Harvard, it still isn't the world. So that contained all the things that I only realized were true when we did the experimental discovery in December '62. It contained the basic ideas of the theory, which his letter had not contained. His letter was very suggestive; it contained the synchronization technique, and the ac as well as the dc Josephson effects, and the Josephson penetration depth and so on. He got some of these things right, which became clear when he corrected me in private correspondence. So I had to reinvent all these ideas for myself, because I didn't have a copy of his thesis until later.

More or less the minute John Rowell said to me he had found it, found the dc Josephson effect, I asked myself, "why isn't it big enough, why is it so small? Why don't we always see it?" I thought about noise, and John said: "There is lots of noise coming down the mains at a place like Bell Labs. So how sensitive is it to noise?" I realized then that there was a coupling energy, and how big the energy was, and that was the straightforward way to derive the whole thing, based on the total Hamiltonian and the effective energy. I realized then that it had been in Josephson's fellowship thesis, but I had never read the thesis, because he hadn't sent it to me yet. And now I looked at it, and I said: "This is all the stuff that I have discovered for myself." I published enough so that we could do experiments. But then I was very careful not to publish a full paper on this, not to publish in full in Phys. Rev. Then in June of '63 I was lecturing at a summer school and wrote it up. (There was an equally obscure lecture in May.) I thought that was sufficiently obscure, that we would not have what the science historians would call the *Matthew effect*, where the big guy gets all the credit. Apparently this is a nonlinear phenomenon. Either the student gets all the credit, or the big guy gets all the credit. In this case, by being very careful I managed to see to it that the student got a lot of the credit. I'm happy with the way it worked out. The only thing is, I think John Rowell should have more credit as the actual discoverer of the dc Josephson effect.

John had these low impedance junctions and Giaever had higher impedance junctions, so it might be hard for him to see it. And there is no reason why GE should have less noise than Bell Labs.[1] In our paper we said that other people may have seen it. John built a shielded room for tunnel junction work after we realized this

---

[1] See Chap. 6. Giaever did publish observations showing the dc Josephson effect, but he makes no claim on the discovery because he did not understand it. This was before Josephson's paper.

noise effect. Before that it was only because he had such great tunnel junctions that he could see it at all in unshielded rooms.

While I was in Cambridge I can remember going up to Birmingham talking about the other Morel Anderson paper, and Rudolph Peierls was in the audience. And after my talk he said: "You say there is a structure in the gap function. Why don't you see it in the experiment?" My first reaction was "Oops!" And then I said, "I suppose we could, in tunnelling experiments." And then two weeks later John Rowell called me from the US, and said, "we have this fantastic structure in lead." And then when I got home, John and I went down to Philadelphia, because I also had got a paper from Bob Schrieffer, about using the online computer, some of the first serious online calculations. He had done these calculations on the Eliashberg equations. He had no idea what to expect, what to put in. We said to him, having looked at Brockhouse's results on phonons in lead, that there are these two peaks, there is a transverse peak and the longitudinal peak. Several papers followed. Scalapino and I also wrote a paper, where we tried to identify the phonon structure. I think we probably did. And then I handed the problem over to McMillan. He did a lot. Sadly he died, before the discovery of high-T_c.

During my long stay in Cambridge '67–'75 I didn't do so much in superconductivity until finally He3 came along. I had a student, David Lancashire, who was interested in high frequency phonons and size effect. And we were doing very delicate phonon transmission experiments of various kinds. Back at Bell Labs, once we had the Josephson effect, Ali Dayem and I were stimulated by some misinterpretation by Parks of what he was seeing with little metallic bridges. This was after Josephson. So we invented the concept of weak link and, incidentally, we patented the whole thing, everything that had to do with a contact as opposed to a tunnel junction. This is how many superconducting devices work, so they are using our patent, by John Rowell, Ali Dayem and myself and owned by Bell Labs. But the Bell labs refused to enforce it in any way. Otherwise I was not doing so much superconductivity at Cambridge.

8.2.11 *Kondo Effect and the Renormalization Group*

I was busy doing the X-ray edge problems and the Kondo problem. We invented or reinvented the renormalization group. There are two kinds of uses of the renormalization group, one is in statistical mechanics, with the accompanying concepts of relevance and irrelevance, but before that the renormalization group was simply a method in field theory, which you use when you have a one parameter field theory problem as in QED. Kadanoff and Fisher had developed the concepts of relevance and irrelevance, and universality. And Wilson hooked renormalization on those concepts, and that was his great achievement. His great achievement was using renormalization group properly in statistical mechanics. But it was not the first use of the renormalization group in condensed matter physics. The first use of it there was a sequence of papers by Gideon Yuval and myself, which was submitted three or four months before Wilson. I didn't know about the field theory renormalization group. I had reinvented the field theory renormalization group in connection with solving the Kondo problem. Yuval and I in the summer of '69 had solved the Kondo problem. Our paper was delayed by the referee, and eventually came out about the same time as Wilson's paper. Schrieffer admitted a couple of years later that he was the referee. Yuval and I were impatient, so we published a couple of papers in other journals. So if you look at the submission date of our big renormalization group paper on the Kondo problem, you will see that it was submitted slightly before Wilson's paper. I don't know if Wilson was aware of our work. I can remember John Wilkins, who was at Cornell, saying to me in the summer of 1970 that I ought to drop by and see Ken Wilson, who was working on renormalization. So we talked and he told me of the renormalization group work and I told him about the Kondo problem, so there was no particular question about the communication between us. (Note added by PWA: I am not claiming that our work had anywhere near the impact or range of Wilson's work, just that he has unfairly ignored it in his own work on the Kondo problem.)

8.2.12 Resonance Valence Bond (RVB) Theory[2]

The RVB came rather quickly in 1987. I had some of the ingredients already. I had this idea of some kind of disordered state made up of singlets. I simply asked the question, are there two dimensional arrays that have this kind of structure? Then I heard Maurice Rice talking at Bangalore on doing the one dimensional Heisenberg model by projective RVB, a Gutzwiller projection of the BCS. Perhaps it was a Gutzwiller projection of the Fermi sea? I stayed up all night and worked out that is the same as a Gutzwiller projection of the BCS state, with a peculiar gap. So the next morning I had this idea of a Gutzwiller projection of the BCS state as some sort of RVB. The other ingredient was that the cuprate was a model of a Mott-Hubbard doped insulator. I looked at the crystal structure and I realized that this was just the spin one-half antiferromagnet that we were looking for all these years, a perfect spin antiferromagnet, and maybe RVB? It isn't, it turned out. On the other hand we could dope it, and the doping would destroy the antiferromagnetic structure. And then we would have an RVB. So all of that happened within three days after listening to Maurice Rice about the one-dimensional case. I went home from that meeting and was absolutely convinced I had the solution of high-T_c. I started writing it up, using thus my three months at Caltech as a Fairchild scholar. It got

---

[2]From Wikipedia, on RVB: *The theory states that in copper oxide lattices, electrons from neighbouring copper atoms interact to form a valence bond, which locks them in place. However, with doping, these electrons can act as mobile cooper pairs and are able to superconduct. Anderson observed in his 1987 paper that the origins of superconductivity in doped cuprates was in the Mott insulator nature of crystalline copper oxide. RVB builds on the Hubbard and t-J models used in the study of strongly correlated materials.*

tougher and tougher to really pin it down. It wasn't obvious why this was true, and it wasn't obvious why the transition temperature was so low, or what shape the theory would take. Baskaran was visiting Princeton, and so I was on the phone back and forth a lot to discuss with him. He was perhaps the first to suggest that RVB in the superconductor had to have phase coherence. It depends on the phase stiffness which is small, so we said this in words in the paper that the transition could be low due to lack of phase coherence. So this diagram that various other people claimed in '91 and in '95, and later, that "here is the RVB transition and here is the real transition which is lower," that was really on our mind. Baskaran and I had published that in several papers. That was essentially right, and I wish we hadn't published anything more, because it was a lot of other stuff that got out of hand.

8.2.13 The Situation as Seen in 2001

As far as the science is concerned, it has become increasingly clear to me that the original insights had the right physics, some of the right physics. It has to be realized that this intermediate state between superconductor and antiferromagnetic insulator isn't thermodynamically stable. And so it can become inhomogeneous in various ways, and you can have stripes and you can have a dirty mess. In a paper I say that stripes are inevitable, they are a disease on superconductivity. The other thing is that it is a formidable problem for the formal theory. There are no small parameters lying around easy. There are problems that are simpler than this, that remain unsolved. For instance, let's take the infinite U Hubbard model. There is no formal theory for when it ceases to be a ferromagnet. There are good estimates for when it is ferromagnetic at low doping. It is a metal at low doping, not a superconductor.

Regarding the idea of a hidden (now called "competing") order parameter, I can't believe it is intellectually serious. It is partly there for personal and sociological reasons, and partly for intellectual reasons which I don't believe are correct. I introduced in one of the first papers the "Anderson murder mystery theorem." The theorem is that if someone murders all the great chefs—you know there is a movie like that, where someone has murdered all the great chefs—you have to believe it is the same guy doing all the murders. So if you have a lot of things happening at the same time, you have to believe it is the same physics. If you have antiferromagnetism and a Mott transition and you have superconductivity right up close to it, superconductivity of a kind you have not seen anywhere else, you have to believe that the culprit is the same. So why I think it is intellectually unacceptable is that it is a Mott-Hubbard insulator, and that is a very difficult system to do formal theory on, but you can be reasonably certain that the fact that it is a Mott-Hubbard system is the key. That is really all that is involved. I guess the only difference between 1987 and now is that it can't be an s-wave superconductor. It has to be d-wave superconductor. But there will always be room for people to say but, but, ... but you can't prove it.

8.2.14 The Future, as Seen in 2001

I think it will take ten years for all the nonsense to die down, for all the emotions that have been aroused by various groups to die down. When all the emotions have died down someone will come along and say, "Baskaran and company had it, really." But there are still questions and there are many, many fascinating phenomena. There is a paper by Shraiman and Siggia which is probably closest to a theory on an antiferromagnet. It disappeared because too many papers were written on competing theories that look much simpler, but are just wrong. Schraiman and Siggia were more or less right. Probably they didn't get the whole story, but they got the crucial thing about it, saying, "this hole is going to be surrounded by a great big soliton in the magnetic structure." That's right, and they have found experimental evidence for that. Sometimes when I have solved something exactly, people say it can't be right because it doesn't agree with conventional wisdom.

When I am asked if I feel that the world is against me in this matter, my answer is that "of course I am paranoid!" Everyone has gone off..., I went off, 1988 to 1998, in a crazy direction. I should have realized it was a crazy direction. I should immediately have accepted that it was a d-wave when they first saw it in the experiment. (Note added by PWA in 2012: Yes, until 1998 I was working on a crazy, deluded hypothesis suggested by some of the many wrong early experiments; this interview was during my rethinking period, and I seem to have had a Shraiman-Siggia phase, now over, not that it's a bad paper, just mostly irrelevant.) From 2001 to 2011, once I returned to my original idea of a Gutzwiller-projected pair function, I solved most—not all—of the cuprate problem.

Note Added in 2012

Due to the recent great interest in the Higgs mechanism we asked Phil Anderson in August 2012 to clarify his role in the story of the Higgs mechanism:

The written reply from Anderson:

"Our interview took place in 2001 before all the recent fuss about the Higgs, and my 1963 paper was rather little known at the time—although it has had rather a faithful cohort of advocates over the years, of which I should mention with gratitude Roman Jackiw and especially Peter Higgs. I did, in the interview, mention briefly the gauge invariance problem of the BCS-theory, on which I published three papers in 1958, the third of which contains the "Anderson-Higgs" mechanism: but I didn't mention the sequel.

The "Gauge problem" was that the current response comes out of the BCS theory in the disturbingly non-gauge-invariant form $J = (\text{const})A$, not $(\text{curl})J = (\text{const})H$, $H = (\text{curl})A$. The cause of this is that the assumed wave function in the BCS-theory is a "spontaneously-broken-symmetry" state, a ground state of a macroscopic quantum system which does not have the symmetry of the underlying Hamiltonian which determines the laws of motion. Such states are fairly conventional in the quantum

8.2 His Own Story

physics of large assemblages of atoms, "solid state physics" (most solids are crystalline) and a good example is the quantum antiferromagnet, which I had studied early in my career. In the course of this work, which mostly had to do with collective excitations of the spins, or spin waves, I came across what seemed to be a very general result, the reasoning for which was borrowed by the particle theorists and named the "Goldstone theorem," with phonons and spin waves as examples: that if the broken symmetry is a continuous one such as spin rotation symmetry, one of the branches of the collective excitation spectrum will have a vanishing energy at 0 momentum. The reason is that in order to restore the true isotropic symmetry in the exact ground state, the zero-point amplitude of this particular excitation must diverge.

The little flurry of activity stimulated by Gregor Wenzel's skepticism of the BCS-theory had two major consequences for quantum field theory. Both Nambu and I developed spinor representations of BCS and used them to "demonstrate" gauge invariance of the theory. But this captured the interest of Nambu in BCS and stimulated him to propose, in 1960, with Jona-Lasinio, that the "vacuum" of field theory might in reality be a spontaneously broken symmetry state. The symmetry which he broke in this theory was conservation of chirality, which is already weakly broken by the chiral anomaly; and the "energy gap" produced is the mass of the hadrons (note that the often publicized claim that ALL mass comes from Higgs is far from accurate). Also, the Goldstone boson of the theory is the pion, by no means totally massless because of the chiral anomaly.

I remained dissatisfied with my gauge paper, particularly because there was no experimental evidence that the gap in real superconductors was actually violated by Goldstone bosons of any kind. It seemed to me that they would, if there, be quite easy to spot. So this makes this another example of my general rule that deep theoretical results often come from experimental anomalies. I seem to have been quite excited about the whole subject because the Letter announcing that I had a whole new method and previous work wasn't right appeared in the same issue of Physical Review as the first paper itself—one submitted in January '58, one in May! The whole paper on the method, then, was Phys. Rev. **112**, 1900, (1958) and is a classic if I ever wrote one—though I myself failed to cite it in the mass paper! I remember working very hard on it. It makes, among others, the essential point that a neutral Fermi gas would have Goldstone bosons, but that in the charged gas which we really have there are none, because they are replaced by massive bosons which come from the three polarizations of plasma oscillations.

I thought Nambu understood what I had done—of course I sent him all the relevant preprints—but in fact he had not, though he still claims that his paper solves the gauge problem. I happened to be in Moscow in December 1958 and had a brief conversation with Shirkov of Bogoliubov and Shirkov, interrupted amusingly by the KGB, and I think B&S did understand. Certainly the Landau group did but they thought Gorkov's Green's function treatment of BCS also solved it about the same time—I don't know if that's right.

Anyhow, when I began to hear that the Goldstone theorem was causing trouble I pricked up my ears. How did I hear? This was the year 1961–62 when a lot happened, e.g. the Josephson effect; I spent a year as Mott's guest in the Cavendish, and

I had a few contacts with the particle world there, particularly Richard Eden and some with the grad student grapevine; Weinberg was around though I don't remember any direct contact. When I got back to Bell I quizzed John Klauder and he had a visitor, John G. Taylor, who was very helpful. (Klauder is one of Bell's anomalies found useful for no obvious reason.) Anyhow, I wrote it up, and rereading it I think it's more complete than it is given credit for. Bob Brout was a fairly close friend during the late 1950s, he's referred to all over my papers, and even may have been a contact about Goldstone. The Kibble et al. group I had no contact with.

The A-H mechanism has had a bit of play recently. In the original paper you will find me remarking that it is extraordinarily hard to find any physical way to measure the presence of superconductivity INSIDE the superconductor. This is an effect of Anderson-Higgs, as Sondhi et al. wrote in Annals of Physics a few years ago—there is no Higgs field, no "order parameter," no condensate for a superconductor as for the Higgs. The electrons which carry the supercurrent are not altered by it, essentially. The technical term invented for this situation is "B-F Theory" for reasons you don't need to know."

Chapter 9
Pierre-Gilles de Gennes: The Orsay Group on Superconductivity

A pioneer in superconductivity and soft matter, an ambassador of science.

Fig. 9.1 Pierre-Gilles de Gennes

9.1 Biographical Notes

Pierre-Gilles de Gennes (1932–2007) was awarded the Nobel Prize in Physics for 1991 *"for discovering that methods developed for studying order phenomena in simple systems can be generalized to more complex forms of matter, in particular to liquid crystals and polymers."*

de Gennes was an unusual physicist in several ways. He is remembered for his creative leadership when setting up mixed research groups of experimentalists and theorists. His mastery of diverse areas of physics, based on common principles, earned him the characterisation as "the Newton of our time." He was the undisputed champion of soft, complex matter, i.e. aggregates of fluid matter, which retains some characteristics of solids. His main contribution to physics was in finding ways to understand the ordering phenomena taking place in seemingly disordered matter, in introducing the necessary new concepts, inventing the necessary language and tools, and in elucidating their dynamics. He addressed wide audiences in person like perhaps no other physicist in history. He is remembered as well for his unique style in communicating with those audiences.

de Gennes got his university education at École Normale Supérieure, Paris 1951–55, and his PhD in 1957 at the French Comissariat à l'Energie Atomique (CEA) on ordering in magnetic systems. He spent some time as an engineer at FEC. He served in the French navy for over two years, and worked for a time with Charles Kittel at Berkeley. He set up the famous Orsay group on superconductivity in 1961. This led to his first textbook in 1966, *Superconductivity of metals and alloys*, still in use today. In 1967 he established the Orsay Liquid Crystal Group. He studied classes of liquid crystals like nematic and smectic phases, and elucidated the physics of their optical properties, thus making an important contribution towards their present-day widespread technical applications. He summed up the field in the famous book, *The physics of liquid crystals*, in 1974. de Gennes was appointed to the prestigious Collège de France in 1971. Later he worked on polymers in collaboration with the Strasbourg group, studying their dynamic and ordering properties. He introduced a number of new concepts in this field, and showed how scaling arguments and scaling functions could be used to describe the ordering properties of solutions of long, entangled polymer chains. Also here he wrote a famous textbook: *Scaling Concepts in Polymer Physics* in 1979.

Later, from about 1980, he worked on colloids, granular matter, adhesion, and wetting. During his last years, which he spent at the Curie Institute in Paris, he worked on cellular adhesion and memory formation. de Gennes had an unusual ability to find the essential parameters of the problem he worked on, and to transform them to simple ideas and theory. In every new field he drew on experiences from previous work. Ideas from superconductivity were transferred to new insights in liquid crystals. Scaling concepts used in critical phenomena in solids and liquids were exploited in soft matter, like liquid crystals and polymers in solution, often referred to as complex liquids. de Gennes' interests spanned very wide, including industrial problems, where he, as an example, made important contributions to the use of surfactants in oil recovery. He has rightfully been called an ambassador of science. The following excerpts from his own account sheds some light on how and why he became the great scientist we have known.

9.2 His Own Story

9.2.1 Early Days

My story is a little bit anomalous because of war conditions and health problems. I never went to school until a relatively old age, until the age of twelve or so. The reason was that we had to travel from Paris to the mountains, because I had poor lungs. When also the German occupation had taken most of France, we stayed there, and for a long time my mother just taught me what she knew, which was mainly literature and history. She was very fond of history, and didn't know much about science. In her days it had not been taught. So I didn't learn anything about science at that time. When finally I went to some form of high school they didn't want me originally because I was too young. But ultimately one high school decided to have me pass an exam. I read for the exam, I went in and I was very happy because we had brilliant teachers, who often had migrated themselves from Paris.

This all happened in a little place called Barcelonnette in the south Alps. And for all that time I liked science but I had no special push. Certainly, the family tradition was to go into medical studies, and my mother's liking was that I would do something in diplomacy or something of that sort. But I resisted that at a later age when I was getting out of high school. I think the reason why I moved to science was really related to some sort of precision test. I mean in our trade you say something, you may advance some very bold assumptions and so on for some time. But after a while you check it and you know if it works. And if it doesn't work, there is something very clear about what you have done or failed to do. While in all these artistic occupations I didn't find the same thing, it was very fuzzy.

Medicine might have been a natural choice due to family history, but I didn't think much about medicine for some reason. My father was a doctor, but he had passed away at that time, so I didn't have a very close example. Above all my grandfather, my uncle, my father and now my son are doctors, but at the time I didn't think much of it, so I think it was rather a question of no information.

9.2.2 Education

I started thinking about science only at the end of high school. I had excellent advice. In the French system we have these preparation classes after high school, for two years, which brings you to what is pompously called *Grandes Écoles*, a place like this where we are now. And this was the natural training, a natural path for us to go through. But some lady at my high school who was, I think the directors secretary—she was married to a professor, so she knew a lot—told me there is a very special system, one particular preparatory class, were you can learn physics, chemistry and biology together. It was called NSE, Normale Science Experimentale.

So I went there and it turned out to be a remarkable form of teaching, much more stimulating then the classical prep class which had very strong math, and very strong formal physics, things like thermodynamics in detail and so on. This was very different. I think one major difference was that it taught us to look at things, to look at plants, leaves and animals; to look for them, fetch them, catch them, and bring them back, for instance insects, put remedies like paraffin and droplet to block them, and open them with two needles and look what is inside. That I found was the best education. I was now back in Paris. There was one class of this type for boys, and there was one class for women at the neighbouring university. These were state run schools.

I had these two years which were a delight, in particular not only working, they had us doing much more experiments than they do in the classical French system. For instance in those days ... I wouldn't be able to do it now, setting it right ... setting up an interferometer, something like this, setting it so that it works. I couldn't do that now! It was some sort of experimental system, different from the typical prep class, and the only one in the country.

And even now, about fifty years later, we make a very closely linked group. For instance we had a very widely known minister of science a few years ago, Claude

Allegre, who was a geophysicist, and who had great programs, but great difficulties and many fights, but he was very well known for the public. Claude Allegre was in the same class, I think one or two years after me. He wrote many books, and one of his books is dedicated to the memory of this class.

9.2.3 Teachers and Masters

This lasted for two years, and then I entered the classical system, which was what we called École Normale, not even a mile from my present office. Originally, it was dedicated to educate high school teachers, in the 19th century, and then it became this centre to educate research personnel and university lecturers. What was nice at the École Normale was that we had enormous freedom. We worked for four years, but essentially we did what we wanted. Very little... We had some great masters, one was Alfred Kastler in physics, the man who got his Nobel Prize for optical pumping. He was a very charming man and a very good lecturer.

And we had another one, who is less known in the scientific community, who is very important, and that is nuclear physicist Yves Rocard. Rocard had worked in acoustics, things like stability of bridges and things of that sort. He had worked in the early days of radios, designing tubes, and was solving electrostatic problems for tubes and so on. And he had also been interested in radio waves during the war. He had been very active in the resistance in France, informing the British about building of rocket systems and radar systems, which made him an admiral after the war, since he had so much served the navy.

So he was a man of many, many interests, and he taught us things like basic physics, electromagnetism, some formulas in mechanics, shockwaves, things like this, and some statistical physics at a very modest level. But he had this enormous appreciation of the field. He had also installed, for instance, a detection system for underground explosions, which was used all over the world. He had very simple and clever detectors. So he was a very inspiring person. A man for all seasons, you might say. His son became the prime minister some 15 years ago, so it's an interesting family.

The third man whom I would mention was Pierre Aigrain, who started semiconductor science in France, and he was lovely. I worked one year with him when I was in *École Normale*. You know, we had one year of experimentation in the lab, and he would come every morning with a new idea and describe it to us, and we would be really open eyes, like this! Sometimes the idea turned out to be completely crazy, but you know, after ten ideas like this, one idea turned out to be a very novel physical effect. And since he had one every day, it turned out to be something like three every month, very impressive! He was extremely nice, he really treated us like friends, not like students. He even brought us—we where mainly two of us this year, it was Philipe Nozières and I—he brought us to conferences all over Europe. He was a remarkable master.

So these three people, Kastler, Rocard and Aigrain were the people who taught us physics in some sense. And I would say sometimes—maybe something stupid

9.2 His Own Story

or pretentious—I would say the fourth man who taught me physics I never met, but by reading his papers, was Feynman. When I was still at *École Normale* precisely working with Rocard, I got interested in the Feynman papers on helium. He had two beautiful papers, with very few equations in them, on the nature of superfluid helium, the wave function of the ground state and rotons. He had this very simple trick to derive the roton spectrum, and I found that absolutely beautiful. I think this really taught me physics. I read Feynman in '53–'54. Something like this. And this was a pure delight. And I'm very sorry I never met Feynman. I was in California at one time. He was in Southern California. We could have met, but it never happened. I really regret it very much. And especially because in later work, for some of the things in the papers I have here on my desk, we used a lot of path integrals. So it would have been very natural to talk to him about it.

9.2.4 PhD

After the École Normale I had a hesitation between three directors, so to say. One was Jacques Friedel, who was starting a group on electron physics in metals, a very nice person. And let me remember ... another one was Anatole Abragam who was starting a group on nuclear resonance. And the third was André Herpin who had a group on solid state and statistical physics. And I went to Rocard to ask his advice about this choice. He sent me, he pushed me towards Herpin, because he had known Herpin precisely working on microscopic structure of shockwaves, within one mean free path, and things like that. He knew Herpin and he liked Herpin. So I went to Herpin. Well, in fact it was a rather random process, because Rocard didn't know enough about modern physics to make a very wise choice. But the funny thing is, that in these doctoral years, actually I worked with three of them. I was under Herpin's guidance formally, but Herpin was a really relaxed person so he let me do what I wanted. For one year I think I did very little.

This was at Saclay. But Morgan had an office nearby and began to send me little theoretical problems at times. And Friedel was a consultant at Saclay, so I started to meet him and discuss things with him too. So I was very lucky, I didn't have one master, I had three masters, and three really good ones. A very happy time.

9.2.5 Becoming a Theorist

I think quite early it became clear that I should be a theorist. I will tell you why: Because I'm very clumsy. At some moment, once we were discussing with Roman Smolukovskii, son of the great Smolukovskii, about some radiation damage problems, which is of interest for the atomic energy. Roman was teaching some sort of group of lectures on that. He was working in Princeton at the time.

And they decided that they would put me in one experiment where I would irradiate some tungsten wires on the van de Graaff, and measure resistivity and things like this, a very simple experiment, nothing sophisticated. But immediately I showed my inability to it, because I managed, among other things, to put my eye in the beam, and then there was panic. Fortunately, nothing happened, but there was a panic, and we all decided that it was better to keep me out. Again, very nice.

So I had a purely theoretical PhD which was mainly on magnetic materials, thinking about structures of pair correlations, in space and time. And at this time the notion of pair correlations at different times $S_i(0)S_j(t)$ was very recent. Van Hove had introduced it in connection with elastic scattering, but it was interesting to think about it, construct it. And then I began to construct these pair correlations, in particular the paramagnetic phase, where it's a fact that you have some form of diffusion problem. But in early times it was not so easy to construct these correlation functions, so I could construct them by methods of moments and things like this. It was an interesting aspect, because I could work in close cooperation with neutron people at Saclay who were working on critical scattering.

9.2.6 Experimental Approach

In my early training I had been taught mathematics reasonably, I mean much more than American students. But it was not the only focus. When I look back I've never used any sophisticated mathematics. Sometimes I've missed things. I can give one example which is many, many years later. I was working at the time on the liquid crystals and I knew also about vortices and superconductivity, and so on. I knew precisely that Feynman had written something very early on the nature of singularities in superfluid helium.

I was wondering why you have walls in certain systems and lines in others, and points, and I was groping at making a classification of this, but I didn't have the right mathematical education. And once I was out picking berries near Orsay. I was living in Orsay at the time and was picking blackberries with Barry Mazur, who is a mathematician from Harvard, and with both wives. You know when you pick berries you have enough freedom to speak, so I was explaining the problem and Barry said, "You know there is a tool, which is very clever for these things, and it's called homotopic groups. You circle around the singularity and you compare what you have obtained by continuous transformation. This generates the group. You should look at this." And I said; "Ha, ha!" and I never went to look at it!

And this was exactly ... some two or three years later, Toulouse and Kleman came and used homotopic groups cleverly and got a complete classification of the singularities in fields. So that's a good example of how I've been weak, or lazy, if you will.

9.2.7 BCS and the Orsay Group

What I had very early was BCS, because I had heard Bardeen just before BCS in Les Houches, describing the existence of the gap. I had it from the horse's mouth these things. I read all the BCS literature, and the Russian literature about propagators, and things like the Ginzburg-Landau equations. After Berkeley I went for 27 months to the French navy to serve, and after that I came to Orsay. While I was in the navy, I managed to keep an eye on the literature. So I had followed that and it looked really stimulating. In spite of this I had not been directly preparing to enter superconductivity at the time, but collecting the knowledge. It was very educational. These papers were beautiful, the BCS on one side, and Gorkov's work on the Russian side.

I have very fond memories from Orsay. The story of this thing is really nice. I went to Orsay as a young lecturer. And in Orsay I was very interested in superconductivity. I decided to establish some sort of little group, and immediately I had some nice theorists showing up, very young people. But I really wanted to have experiments too. And I got the help of some colleagues who were very nice and helped me, for instance to find rooms, which was not obvious at all. Aigrain and others helped me there, and some helped me to install liquid helium and tubes because we needed to recuperate the helium. At the time helium was rather precious. So we set up two or three rooms. One was on the top of the roof. It was completely illegal, we had no right to construct it there. But we got it, and I got a lab going.

I was not involved in doing the experiments, I was just suggesting them. I had something like four young PhD students, plus one post doc who had not really worked in superconductivity but had worked on magnetism at low temperatures. So he knew the low temperature techniques, and one engineer, and I think it was something like three young PhD's for theory. And we started with that. Originally I was really an enthusiast. Then one year passed, and not one experiment worked! So I had these four students and it didn't work at all for a time. And we are saved, the experiments are saved by a theory in a very strange way. What I mean is the following: I had among my students Étienne Guyon, who worked with me on a little theoretical paper, which maybe is in this pile on my desk somewhere, on proximity effects, I think in dirty materials. I don't remember exactly what it was about.

But it was a very simple thing and indicated the effects of proximity in simple terms. It was easy to read. Then he goes to Cambridge, England, for some reason, and he mentions this thing. And the people in superconductivity which were around, Pippard and others, liked it and talked to him, and he is accepted, so to say. But then he says, "We were trying to make tunnel junctions, but our junctions never work. We evaporate, and we never get the right alumina layer." And they tell him "Of course, you have not done what you should do, which is not written in the published literature, which is to have a plasma discharge when you put the alumina on."

He goes back with the message, and the next week we get our junctions working, and from that we got four PhD's, some on tunnelling excitations. All these papers here on the table in front of me are on structures of excitations, and some on H_{c3}

and with particular field properties. So with that my problem was solved, these kids were saved and had good PhD's!

And in fact, it's interesting to see what we became. Étienne Guyon had a very diverse career. After the superconducting period he went into liquid crystals with me, and he invented some beautiful instabilities, hydrodynamic instabilities in liquid crystals, and then he went into granular matter, random media, porous media, percolation and so on. He wrote books about this. Well, he became the director of precisely this École Normale, where we both had been in our youth, and now he is retired and he is an emeritus professor here. So he has had a very amusing trajectory.

It was very bold for an experimentalist like him to move from for instance superconductivity to liquid crystals. For a theorist it is not much cost, you just bring a piece of paper and a pencil. But for an experimentalist it's a different world. You don't know the people, you don't know the language, you're not accepted easily. It's a very long patience thing. The instruments are different and you don't know how to work with them in the beginning. But I had luck in some cases. Étienne is an example. I have a few examples of people who were very bold and were able to do it. And in fact, Étienne brought some techniques from low temperature. He knew about evaporation, and he did some very useful surface treatments with angular evaporation, were you set the beam at an angle on the surface. This creates anisotropy in the surface, and that could be applied to nematic liquid crystals too. These things have become quite important, and he started it. Also, in thermal conductivity he knew a lot about thermal transport and he could measure thermal features of liquid crystals with little evaporated things, very small samples, beautifully. So the techniques of low temperature turned out to be useful in this complete room temperature sort of thing.

9.2.8 Liquid Crystals

Going into liquid crystals I could exploit certain similarities and experiences with superconductors and superfluids. There was a little theoretical aspect, for instance in that smectics are characterized by two-dimensional sheets of liquid, and you can describe them with an amplitude which says, "Am I in the sheet region or in between sheets." And the phase which says how you shift between sheets, so it is a bit like a superfluid. And when you look at the defects for instance, the analogue of vortices are certain dislocations in the smectics. And there are some interesting similarities like phase. And really, two people jumped on that at one time. One was a great man who we miss, Bill McMillan, working at, I think Urbana, at the time. We perceived this. Well I perceived parts of it, Bill was more interested in the microscopic properties on the Angstrom scale. I was more interested in textures, defects and collective mode, long wavelength.

The collective mode analogy was also very interesting, it was an analogy to some sort of zero sound, we called it, and so on. So it is true that the low temperature education was helpful at times, in this question.

9.2.9 Main Achievements at Orsay

As far as the main achievements at Orsay is concerned, one side was excitations; we understood a number of things about excitations. For instance excitations, tunnelling and phase in type 2 materials, where there is no gap and there is still superconductivity. This was very important for us to prove, that you could have superfluidity without the gap. There had been all this folklore that the gap had been the source of superfluidity, and it's not. So there was all this excitation work in Orsay, then some Ginzburg-Landau work, and H_{c3} was one example, and also some more delicate properties related to for instance proximity effects where we have a strong conductor, something like copper. When put in a magnetic field it will pick it up at very low fields. The superconductivity leaks into the copper. But when we increase the field, you suddenly jump and the copper gets completely killed. You jump as the field penetrates.

The man who did this was an interesting case, Strobol Duro was his name, he was a PhD student, and we had him—which was pretty bold too—part in theory and part in experiment. And low temperature experiments are not very easy to have enough time to do, but he was very effective, and he did that. He also worked to some extent with liquid crystals later, and when he went to work with Phillips. One time, for a short period he was even the global director of research for Holland Hoven community. So he basically had a beautiful career, after that I think he suffered from problems, but I mean, in a big company like this, at times you are getting difficulties with people fighting each other. He had lots of problems operating that place. But he was an interesting case and he had started with that sort of thing. So to come back to the question of achievements at Orsay, I would say one aspect was structural excitations in the sense of fermion excitation, since we also calculated excitations near the vortex in a pure system. The other was these Ginzburg-Landau features like H_{c3}, all these more delicate features with penetration of fields and mixed systems and so on. And some ideas of the collective modes of vortices too; how you can drive them by Q-waves and so on. We also did a bit about co-existence of magnetism and superconductivity.

It's amusing since it was early and it didn't go very far at that time, but now we have epitaxial films, very thin, very controlled, we are beginning to see things. I have some beautiful experiments were you diffuse layers into a magnetic medium and you still find some coherence.

9.2.10 Why Leave Superconductivity?

I have met this situation a few times. There is a moment; I had worked for the first wave with something like four PhD's in experiments and also three or so, three or four in theory. I had young people who were very brilliant, Étienne Guyon, which I mentioned, but also Guy Deutscher, who is now in Tel Aviv. People like them. And I had the feeling that they could do at least as well, probably a little bit better then

I would do, in front of any new problem. And so I think it's very much like with children, there is a moment when you must be independent, and let them do their things in peace. And if they ask you something, of course we can help, but try to move on, and think that is a very healthy system, it gives them responsibilities early enough.

Because in France, in many old Latin countries, maybe also in many other countries, we tend to have the opposite trend where a big chief keeps the power over a pyramid for many, many years, and that's not very healthy. So I had the feeling, for the first time there. And at that time we had one seminar by a French pioneer in polymer science, Charles Sadron. He gave us one seminar on polymers in general, and I found this absolutely beautiful. I said, "Let's go to polymers!"

We tried, the theorists, not the experimentalists. We tried for two or three years, I think, and we produced some ideas, which are still a bit useful, but I was not happy because we did not have an experimental group. There was a strong experimental centre in Strasbourg, where it is still now, run by Sadron. And they welcomed us very kindly, they let us enter the field. But nothing in response, so after two or three years I decided that this was too dry, that we were just cranking theory but not achieving real life, or something. So then we went into liquid crystals, where we had some four or five years, maybe six years.

You might think we were frustrated with progress in superconductivity. But it's a bit different. I'm sure that my generation has had this feeling a little bit later. We had the impression that T_c was saturating. But at the time there were also some technical points. In the following sense; if we wanted to do new things we felt that we had to do a lot of metallurgy, with precipitates to pin the vortices and things like this. And this was important, I could see it, because I worked very much with General Electric, they could do that, had the means to do it.

But we were a very small group, and we were used to use the simplest alloys, things like lead, tin, things like this that are very easy to make, no problem with metallurgy. Our evaporations were very simple. So I had in mind that we couldn't do the next step. That was another important reason.

There was one experiment that was done at the atomic energy centre, with niobium, and that is a little bit more difficult to manipulate. This was the notion of seeing the vortex lines by neutron diffraction, and you needed niobium to have a strong superconductor and a relatively large field, so you had something to measure. This experiment was very amusing. I remember describing the "to-be-experiment" in Bell Labs once, which was probably around '63–'64 or so. And I said, "Let's use a variable magnetic field. Since neutrons have magnetic moments there is a scattering amplitude which should give about a millibarn or something. There is not much scattering. But maybe you can get a signal out of it." The Bell Labs people at the time were kings of solid state. And I remember Jim Phillips, who I had known far before, said these experiments would never work. And it worked, and I was deeply happy for the Saclay group and the cooperation between my people and Saclay people, it was very nice.

9.2.11 High-T_c

Similar experiments are now often carried out in high-T_c materials. I'm rather ignorant in this area. I don't follow the whole literature. I just talk to my people at times and they brief me on what is happening. But certain strong feelings I do have. Number one, a feeling of apology, because I was of this generation who believed that there would never be any higher T_c's. There has been, let us say, the BCS-Anderson period. But then there have been many attempts by people like Bill Little to add things like organic molecules, to introduce the coupling through electric polarization of the molecule, but it didn't work very well. So I would not have put money into superconducting research. But I admit it fully, I have been wrong in many instances like that.

And when it came, what struck me is the fact that there has been a rush... I see it also in other fields, a rush of theorists who do not really think about the system, but think about the theoretical tool they know, and want to plug it into the system. I see this in biophysics just as well, and many other fields. Everyone has plugged in his own little model, a thing in which he knew how to operate. That usually has been disastrous!

I would particularly say this, and maybe I'm wrong, I may be completely wrong, but all these people who thought about the role of magnetic ions, magnetic modes, spin-wave couplings and so on, people like Pines, Schrieffer, etc. I've never believed much in their theories. This is because when you look at the phase diagram and look at the doping, the region which is the most interesting is the region which is very far from the anti-ferromagnetic limit. I have always been convinced that this is one of these cases where you try to impose a model that you know in reality that it is different. If this is an interesting reality, I'm not sure, but I'll keep two or three ideas.

One idea is that the Jahn-Teller effect may be important. And that was in fact Alex Müller's original reason to work. It is not clear if it works or not, but it's an interesting notion.

Another idea which is related to Jahn-Teller effect is when you have an ion like oxygen which is not in a good minimum but has two close minima, and an electron goes by and has this oxygen jumping, tunnelling between minima. That has very large matrix element, much larger than vibrations which we all think about. It might be an interesting clue.

Then there is the idea of narrow bands. They have peaks in the density of states. We have one man working on it here, Julian Burre. It may be part of the story.

My overall impression is that high-T_c may be more like the one-dimensional superconductors, like the TCNQ family. Originally we were very excited over these systems. One d-wave. A very simple scheme. It turned out it was not at all that simple. It may well turn out that there is a superposition of different effects. TCNQ are very complex, they are not strictly one-dimensional. You are superimposing very different effects.

I have another example, heavy fermions. When they showed up, people were very excited. But they are again a superposition of many types of ordering instabilities.

9.2.12 Popular Lecturing

I've always had in mind that I would like to visit high schools, but originally I couldn't do it because I was the director of this place. And the head of the high schools, and especially prep classes, couldn't accept to have a director to come and lecture, because it would look like propaganda for one of the Grand Écoles. So then we would have to do it for all. So it was impossible originally for me to be invited into high schools. But when I got the Nobel Prize, the change was that the students invited me! Not the directors, and then I could go.

Then it became known, and others wrote to me and I continued. I visited 200 high schools, and it was very exhausting. I spent a good part of two years doing mainly that. Mainly in France, but also somewhat in England and some times also in unexpected places like the Caribbean Islands. It was a great pleasure. And in France I would talk about things like soap films for one hour, and then there may be questions from members of the faculty on technical points, and then we would go on to major points, like, "what should we do if we're interested in science?" and, "what is science for?" and, "what is life?" and, "are there others in the Universe, and what are we doing with it?" But this was extremely nice. And also in these talks I was very critical of the French system of education from bottom to top, including these places where we are now. I'm not a great defender of the French system. It's there, so we have to live with it, but I don't defend it.

I was criticizing all that, and criticizing sometimes programs for teachers, and the teachers were participating in the discussion. So it was rather exciting, but when I stopped, I thought: Well this is a soufflé, you know one of these cakes which collapses. After production it will go down. It did go down for two or three years, but then to my great surprise it was not completely lost. We began to see, coming from the ministry of education, some new lines of operation which were reflecting these discussions. So, ultimately some results came. For instance in the prep classes, which I was talking about, they put in a new thing to do which is some sort of study of an experiment. Each kid has to study one recent experiment, even visiting the lab or looking at recent literature. And from this he will produce some sort of little report. And this has had a lot of impact. Number one, it introduces them to practical science, and number two, the teachers are just as ignorant as the students in front of this new experiment, so the contact between teacher and student has become much better. In French they now say "tu" instead of "vous" and that's a major change in sociology. So these things grow and they have grown a little bit, so I don't regret having done it.

The model can be expanded because it's relatively easy to find some kind of review paper where some experiments are presented rather simply. For example in liquid crystals Françoise Brochard and I once constructed this fine thing which was called "five easy pieces," and it went, I think, into some British journal, and it described experiments you could do with very little means, in liquid crystals. You needed a binocular, not a microscope, but just some sort of strong lens, or something like that, and a little quantity of material, but very small, decigrams at most. And you needed a field, a little magnet, and with this you could do at least five experiments,

9.2 His Own Story

which is very nice. So I think it's very possible to find experiments for young people that are interesting, and not too hard to grab, and sometimes we can make it on the table. Liquid crystals is an easy case, it's easier than usual, because it's visible to the eye. With low temperature physics you cannot show much, it's very hard to visualize. I'm very pleased with what I get out of these things, they are nice, old memories.

Chapter 10
Johannes Georg Bednorz: Discovery of Cuprate Superconductors

"Then I went to see a lab technician apprentice working with me and said, "We have to prepare this compound. Immediately!""

Fig. 10.1 J. Georg Bednorz

10.1 Biographical Notes

Johannes Georg Bednorz shared the Nobel Prize in Physics for 1987 with K. Alexander Müller *"for their important break-through in the discovery of superconductivity in ceramic materials."* Their discovery was one of the greatest news events in the history of science. Several thousand scientists all over the world got involved in research on cuprate superconductors, partly due to the great fun with science involved, partly due to the overwhelming expectations for technical break-throughs. See also Chap. 11.

J. Georg Bednorz was born in Neuenkirchen, Germany in 1950. His fascination with science was awakened not by physics, but by chemistry. He started to study chemistry at the University of Münster in 1968, but ended up with majoring in crystallography. During two periods as a summer student at the IBM Zurich Research Laboratory in Rüschlikon, near Zurich, and later as a diploma student in 1974, he worked under the guidance of Hans Jörg Sheel in the Physics Department headed by K. Alex Müller. His diploma work was on $SrTiO_3$, a speciality of Müller's, who encouraged him to continue his research on perovskite materials towards a PhD, supported by IBM. This is what he did at the Swiss Federal Institute of Technology (ETH) under the combined supervision of Professor Heini Gränicher and Alex Müller. His thesis work was on the crystal growth and solid solutions of perovskite type compounds, investigating structural, dielectric and ferroelectric properties. Upon completion, he joined the IBM lab in Rüschlikon in 1982, where he remained until his recent retirement.

Already while Bednorz was a student at ETH in 1980, Heini Rohrer at the IBM laboratory had asked him if he could prepare crystals of $SrTiO_3$ doped with Nb for the purpose of studying the superconducting properties of this material under varying doping conditions, together with Gerd Binnig. It turned out that the superconducting transition temperature increased by a factor 4. The implication was that the gradient of T_c versus doping was very steep. But when Bednorz joined the IBM laboratory in 1982, this line of research had been stopped, since now Rohrer and Binnig were working on the scanning tunnelling microscope, also to be awarded the Nobel Prize.

In 1983, Alex Müller, having spent two sabbatical years at the IBM laboratory in Yorktown Heights, New York, where he had done work on granular superconducting Al, approached Bednorz again, and asked if he would join him in an attempt to go new ways in superconductivity. The idea was to exploit a polaronic interaction using Jahn-Teller ions, a field championed by Harry Thomas. Müller thought the mechanism might work for superconductivity in perovskites. From then on a systematic effort was made. This was a low cost project carried out as a side effort along with other ongoing management approved projects. The first attempt was to go for classical Jahn-Teller systems like the lanthanum nickelates. Here, La was replaced by Y.

After one year the project was in danger, since the results were discouraging: All compounds turned out to be insulators. Bednorz now suggested to use copper instead of nickel to achieve the Jahn-Teller effect. Electrical conduction was obtained, but no superconductivity. Bednorz then discovered publications by the French group of Raveau and coworkers in Caen, on the Ba-La-Cu-O compounds, and realized they should modify the A-position of the ABX_3 instead of the B-position. Already in the first measurement, in January 1986, a dip in the resistivity was found at 11 K. The test for diamagnetism could not be performed until a SQUID magnetometer had been acquired in September. However, the results were stable and reproducible. In the fall of 1986 a Japanese group headed by Tanaka at University of Tokyo announced results that confirmed superconductivity in lanthanate. Their own work also showed the Meissner effect.

From now on superconductivity was a matter of great public interest. A new era had started. Georg Bednorz continued as a scientist at the IBM laboratory in Rüschlikon near Zurich. He is the recipient of numerous awards and prizes. Among these: The Fritz London Memorial Award (1987), the Hewlett-Packard Europhysics Prize (1988), and the APS International Prize for Materials Research (1988).

10.2 His Own Story

10.2.1 Path to a Scientific Career

I was born on May 16, 1950, in Neuenkirchen, Germany, as the fourth child of a refugee family. My parents, originating from Silesia, had lost sight of each other during the turbulences of World War II, and my mother with my sister and my two brothers had to leave home and were moved westwards. So after their happy reunion with my father in 1949, I completed our family as a latecomer.

During my school years, my fascination for natural sciences was aroused—and almost equally important maintained—while I was learning about chemistry. Nevertheless for my later career, I envisaged to study medicine, to become a medical doctor. Unfortunately, the problem was that my school grades were not really fantastic and access to those studies at all universities was limited. I was realistic enough to understand that I would not have been accepted. So when I entered the University of Münster (in North Rhine Westfalia, Germany), I started to study chemistry. But soon I felt lost in an environment designed to handle hundreds of students. Consequently, I looked for an institution with a more personal atmosphere and I was happy to learn that the Institute for Mineralogy and Crystallography was looking for new students. Together with two friends, I made the change and found intensive and personal coaching with a curriculum providing a comprehensive and solid education in Chemistry and Physics.

10.2.2 A Student's Experience of the "Real Word" of Research

In 1972, after two years, I was asked by my professor whether I would be willing to spend some time outside the university and get some real-life experience. He had been asked by Hans Jörg Scheel from the IBM Zurich Research Laboratory, whom he knew from their common time at the ETH Zurich, whether he could recommend a summer student. I felt that this was a unique chance, and immediately agreed. So very soon, during my summer break, I travelled to Switzerland for this new endeavour and to dive into the unknown, the world of research. At the IBM Lab, I joined the Physics Department, which was headed by Alex Müller, and where Hans Jörg Scheel was responsible for the growth of crystals for the other members of the department. Under the guidance of Hans Jörg Scheel, I learned various methods of crystal growth and broadened my knowledge in materials characterization and solid-state chemistry. It was a good experience to see how my university education could be applied in the real and practical research environment. And I was impressed by the freedom that even I—as only a student!—was given to work on my own.

Lack of experience did, however, in some cases lead to mistakes, but by learning from mistakes, I lost the fear of taking risks to approach new problems with my own ideas. After three months, I went back to Germany to continue my studies. Since I had gained substantial experimental experience, I was given the position of a technical student assistant. My task was to collect relevant experimental equipment, and design, set up and test special experiments, which were later run by my fellow colleagues in their advanced experimental courses.

Apparently my contribution as a summer student to the research project at IBM was well appreciated, and therefore I was invited for another four-month stay the following year. And another year later, in 1974, I could stay for half a year to do the experimental part of my diploma work, which I finished in Münster in 1976.

10.2.3 PhD Studies at the ETH and First Encounter with Superconductivity

My diploma work was on the growth and characterization of $SrTiO_3$, which is regarded as the model system for materials with perovskite structure. The IBM Physics group under the leadership of Alex Müller was intensively studying the fundamental properties of perovskite oxides, and, noticing my successful work, Alex asked me whether I would be willing to come to Switzerland for my PhD thesis. So, in 1977 I started my thesis work at the Solid State Physics Laboratory of the ETH Zurich, under the supervision of Heini Gränicher (ETH) and Alex Müller (IBM), and with support from IBM. I started to grow crystals and gained my first experience in low-temperature physics through the experiments to study structural and ferroelectric phase transitions in solid-solution perovskite crystals. And I continued learning about the fascinating properties these compounds have and how they change with doping. I studied the influence of ionic substitution on the structural and ferroelectric phase transitions in strontium titanate, $SrTiO_3$, and determined the low-temperature phase diagram. By adding barium or calcium, I could drastically lower or increase the structural phase transition in $SrTiO_3$, known to occur at 105 K in the pure case. Another important result was that when the doping exceeded a critical concentration, a ferroelectric phase occurred in the 30 K range. This work was published in my first joint paper with Alex Müller.

During my time at the ETH, I also had my first encounter with superconductivity. One day I received a phone call from Heini Rohrer, a manager in the IBM Physics department and my administrative contact to the IBM lab. He told me that they had a new hire in Rüschlikon, Gerd Binnig, who would like to study strontium titanate at low temperatures. Strontium titanate, if reduced, that is, if oxygen is removed from the lattice, turns into a conductor and even superconducts at 0.3 K. The idea was to dope it with niobium to introduce more carriers in this way and eventually to get it to become superconducting at higher temperatures. Could I provide some crystals? Well, I said, in principle, if nature allows it, you will get it. And I immediately prepared the powder mixtures needed, made the crystal growth by flame fusion, and two days later I brought the crystals to the IBM lab. Gerd started first measurements on my crystals and could show that by introducing carriers by means of niobium substitution, the critical temperate, T_c, increased. We varied the carrier concentration by changing the concentration of dopants, and the T_c soon reached

1.2 K with only a small amount of niobium. That was a tremendous change—by a factor of four! We asked ourselves, "with such a steep slope, where would the T_c go if we went to higher doping?" But unfortunately, 1.2 K was the maximum, and on further doping T_c decreased again. This was my first experience with superconductivity, which would later have its consequences. Around the same time, during my PhD thesis, Alex had left for a sabbatical at the IBM Research Center in Yorktown Heights, USA, to work on superconducting granular aluminium. That was his first engagement in the field of superconductivity, and after his return to Zurich he taught introductory courses on superconductivity at the University of Zurich.

10.2.4 New Project

In 1982, I joined the Lab in Rüschlikon as a research staff member to continue to work on insulators and to study ferroelectricity at low temperatures. My collaboration with Gerd Binnig had stopped long ago, because Gerd apparently had lost interest in superconductivity. Very disappointed, I could see him pursuing a new project and recording strange wiggly lines on an X-Y plotter. This turned out to be the first experiments with the scanning tunnelling microscope![1] It was one year later, when Alex took me aside and asked me whether I would be interested in working on a new project in superconductivity. He had learned about the Jahn-Teller polaron model developed by Harry Thomas for the case of intermetallic compounds, and came up with the idea that polarons in conducting oxides could eventually lead to superconductivity at relatively high temperatures. Would I be willing to work on this topic? I said yes without any hesitation. I heard later that he was very surprised because of my immediate positive reaction. But it was my earlier experience with Gerd and niobium strontium titanate that made me immediately agree to go for it.

10.2.5 Risk

The new project with Alex started around 1983 as a side project. I could not afford to devote my full time to an activity that carried the risk of never leading to success. Alex and I had even agreed not to talk to anybody about our idea. So I continued my work on the insulators, now on solid solutions, and doped compounds to study structural phase transitions and ferroelectricity at low temperatures. I then also spent roughly 20 % of my time on the superconductivity effort. The facilities for materials preparation and the characterisation tools were the same as for the transparent compounds, and nobody realized that some of the compounds I worked with were black ceramics. The only big change with the new materials was that I had to start to measure the temperature dependence of the resistivity. I found the necessary equipment

---

[1] For which he and Heini Rohrer were to receive the Nobel Prize a few years later, in 1986.

for this in the Lab's semiconductor group and had an agreement with my colleagues that I could use their equipment after working hours. The materials processing and characterization could be done anytime during the day.

The Jahn-Teller polaron model "directed" us to a specific class of materials, namely conducting perovskite oxides with a Jahn-Teller transition metal ion in an octahedral environment. The material of choice to start with was lanthanum nickel oxide, a metallic conductor with a Jahn-Teller distorted nickel oxygen octahedron. The idea was to modify the metallic bandwidth of the material to make it comparable to the Jahn-Teller stabilization energy. And we started by partially replacing nickel with aluminium.

To cut a long story short, this approach was without success: we only produced new insulators. Also the introduction of internal strain by replacing lanthanum with the smaller yttrium ion failed. Therefore, I dare say that after one year the project was at a stage where it could just peter out because no encouraging signs were visible. We started wondering whether the target we were aiming at existed at all.

10.2.6 Cu-Components

One day I said, "why are we only working on the nickelates? Nickel is not the only Jahn-Teller ion and maybe we have a chance if we tried for instance copper compounds." Using again the lanthanum nickel compound as host compound, I started adding copper to the system, and found that the resistivity changes in this new series were not as dramatic as those observed so far in all other doping attempts. Although almost all compounds we tried had changed to an insulating state at low temperatures, the compounds with copper were different. They remained conducting down to low temperatures. But again, no signs of superconductivity whatsoever. It became hard not to lose confidence in the whole idea.

I had to—not for the first time—give me a break. On these occasions, it was good to sit down and reflect about what I had done in the past. And also, I went to the library by myself, looking for some distraction. Out of curiosity, I leafed through some journals just to see what other people were doing in materials science, and I found an article in the Materials Research Bulletin about barium lanthanum copper oxide (BaLaCuO). It was an article by Raveau and Michel and some colleagues that showed a linear decrease in resistivity with decreasing temperature, like in a metal. And I immediately recognized what we had done wrong until then. It was wrong when we tried to modify the B position, the centre of the octahedron in the ABO_3 structure. I realized what could be done with the lanthanum barium copper compound when the substitution was made at the A position only, without touching the octahedral position. By heterovalent substitution, i.e., by replacing the trivalent lanthanum partially by divalent barium, one could change only the electronic character of the octahedral site, that is, the valence of the copper position, and avoid chemical disorder at this site. I immediately went to the copying machine and made two copies of the article. Then I went to see a lab technician apprentice working with me and said, "We have to prepare this compound. Immediately!"

10.2 His Own Story

The purpose of the Raveau work was to investigate the catalytic properties of oxygen-deficient compounds, and thus the French group was interested in measuring properties at high temperatures. The paper contained a resistance curve as a function of temperature down to minus 100 degree Celsius. It was an extremely nice, straight line, as is observed for metals. However, the measurement stopped at -100 °C; apparently the French team was not interested in the lower-temperature part, because they did not expect anything interesting to happen there.

Ok, so we prepared the first compounds, and then something happened that caused a delay in my work. The director of the IBM Research Division from Yorktown Heights, Ralf Gomory, had announced that he would come to visit the Rüschlikon lab, and I had been chosen to present my activities to him. So I was busy preparing this presentation. It took quite a while, and I was getting more and more nervous as the date of the visit approached. But I think the presentation went well. I was talking about the possibility of modifying the physical properties by substituting different elements in a given compound, and showed examples of ferroelectricity induced in $SrTiO_3$. I did not talk about my side activity, the search for superconductivity, because after almost three years there was nothing concrete to show, only the idea of people apparently on a wrong track.

I dare say that talking about a topic like that, namely, searching for superconductivity in oxide materials, and even at higher temperatures than known at the time, could have damaged our reputation. So, it was wise to avoid an exposure and thus exclude that someone would question our scientific seriousness. But nevertheless, I tried to leave the message with Ralf Gomory that in materials science, pursuing the path of chemical substitutions, one could have dramatic changes, even a complete change of the physics of a compound. He seemed to agree, said ok, nice presentation, thank you so much. And off he went.

10.2.7 Discovery!

I needed two weeks to relax from the stress of that visit, and after that I travelled to South Africa for three weeks of vacation over Christmas and New Year. And when I came back, I asked the apprentice: "Well, where did we stop? What was the last experiment?" And then remembered, "Oh yes, we have this bottle of black stuff here that we prepared in November." And I said, "Well, this needs to be measured immediately. So let's make the sample preparations."

It was January 1986, when I started to prepare pellets of the new black stuff, cut them into small bars with defined dimensions and mount them for resistivity measurements. In the first measurement, the resistivity went down as usual and at around 100 K it went up again, as I had seen so many times before. But at 11 K, there was a sudden drop in resistivity by 50 %. When this showed up, I thought: "Is this it?" But then I started to have some doubts, and measured again, and a third time. The effect was extremely reproducible. By then, it was late in the evening, so I went home with the feeling that we had indeed found what we had been looking for, and had a beer.

Then came a period when I discussed with Alex how we should continue. What kind of substitutions should follow, what concentration range should be covered in the next series of new samples? Soon we could optimize the materials so that they exhibited a complete transition to zero resistance, and increase the onset of the resistance drop to 35 K. We asked ourselves whether there was an explanation for the resistive transition other than a transition to a superconducting state. We did not find any.

But we were in a dilemma. We did not have a magnetometer to verify whether a transition to diamagnetism occurs, the second proof for the presence of superconductivity. Luckily enough, however, we had ordered a SQUID magnetometer at the end of 1985, well prior to the discovery of the first sign of superconductivity. Towards the end of the fiscal year 1985, as usual, there had been some money left in the budget of our Physics department. And Eric Courtens, the head of the physics department at that time, had asked people for proposals on how to spend this money, and I had said: "Well, we have some activities in magnetism in our department, so why don't we buy a SQUID magnetometer?" And being really cautious, I did not tell him at the time that we were looking for superconductivity and that eventually I would also have to do magnetic measurements. So the SQUID magnetometer was ordered. But although we discovered the first transition in January 1986, we would have to wait until September for more proof, until finally the SQUID was delivered.

By April 1986, however, we had collected so much experimental data and gained so much confidence on our interpretation that we decided to publish our findings even without proof of diamagnetism and although our samples were multiphase.

10.2.8 Identification

After the submission of the first paper, we entered a very busy phase. New powders had to be prepared, but more importantly, X-ray analyses were done to identify the chemistry and the structure of the superconducting phase because the way I prepared the powers was completely different from the method Raveau and his team used. As a result, it so happened that we had obtained a multiphase powder, and therefore we tried to separate the pure superconducting phase.

At that point, Alex and I decided we should enlist the help of Masaaki Takashige.

Masaaki, a specialist in ferroelectricity, was a visiting scientist from Japan, whom Alex had invited to spend one year at the IBM lab and work with me on ferroelectric materials. He had arrived at the end of January 1986, exactly one day after I discovered the first sign of the resistive transition. During the first three months of his stay at our Lab, I introduced him to the method of thin-film deposition and helped get his project started, which was the growth of thin ferroelectric films. So now in May, I asked him, whether he would like to join our effort. He was very, very cautious with his comments, even sceptical, when I presented what we had discovered. He was not a specialist in the field, but certainly knew enough about superconductivity to realize that if this was indeed superconductivity, it would go much beyond what specialists expected at the time. I could see that he did not feel very comfortable with the situation of joining a possibly shaky project. But after he got directly involved in the experiments and could see how reproducible everything was, he soon was fully convinced. He got involved in all sample preparation and characterization steps, and we soon were able to separate the pure superconducting phase and determine its layered structure. In September we started our work at the magnetometer, mutually supporting each other, since we did not have any experience with the magnetic characterization of a superconductor. To get used to this method, we took lead as a test sample, and compared the result with the data in a textbook. After the successful detection of a diamagnetic signal, the moment of truth came for the first LaBaCuO sample—and wow!, also these samples were diamagnetic! The Meissner effect was present! This final proof was made at the end of September, and we rushed to get this published. In the following weeks and months of his stay at Zurich, Masaaki was heavily involved in extensively analysing the time dependence of the magnetic response in different samples at different magnetic fields, and also studied irreversibility effects in the magnetic flux flow.

10.2.9 *The Meissner Effect Paper*

Here is a nice story: We were very quick in doing the first measurements on the Meissner effect, wrote the paper and wanted to submit it to Europhysics Letters. This was in October 1986. We had just made the last corrections and completed the clearance (sign-off) form for publication, which had to be signed by Heini Rohrer, who was the department manager. Alex, Masaaki and I were sitting together, and I had the clearance form in my hand when we heard an announcement via the PA system. The announcement said: "Minutes ago our colleagues Heinrich Rohrer and Gerd Binnig were awarded the 1986 Nobel Prize in Physics." We were very enthusiastic at that first moment, but soon Alex's reaction was a little bit funny, and he became very serious. He said: "This paper is of highest importance, and somehow I felt that we had to rush this paper out, but now Heini will not sign anything for the next couple of weeks..."

I said to Alex, "wait a minute—I'll take care of this." And with the clearance sheet I rushed to the reception desk, and waited for Heini. And as he came down the stairs of the main building from the ad-hoc meeting in the director's office after the announcement, I stopped him and said: "Heini, congratulations on the prize, can you give me your autograph as Nobel Laureate?" And I handed the clearance sheet over to him. He gave me his autograph without knowing what he signed. So the paper went out to Europhysics Letters. I was rather proud of the coup I had landed to get the signature.

Even today, I still have a hard time to accept the further handling of our manuscript. Although everybody in the scientific community had realized the importance of our work, this urgent and final proof for superconductivity was delayed for several months: we had submitted our manuscript in October '86, and it appeared finally in February '87.

10.2.10 Reactions

Two months after I had seen the first signs of superconductivity, I was visited by a former colleague from my student days. He was interested in finding out what I was doing. I said that I had been looking for conduction phenomena in oxides, and we had discovered a transition to superconductivity at temperatures 50 % higher than ever obtained in classical metallic compounds. He said: "Yeah, that's indeed interesting." I said: "It may come as a big surprise for the science community, and I think with this work I'm pretty well established in the IBM Lab from now on." So I was convinced that this was a really important contribution in the field of science.

However, we felt that it would take a long time before our results were widely accepted. And this feeling was confirmed when we presented the results of our work to the scientific community. So Alex went to Germany to give a presentation at the University of Regensburg in the fall of 1986, after the publication of our first paper in September. I went to the University of Berlin and had a presentation in which I included the Meissner effect as the final proof of superconductivity. In both cases, the reaction to our results was almost non-existent.

But this changed dramatically in November. Masaaki Takashige and his wife had come to visit me and my wife at home for a dinner. During the dinner, Takashige's wife Emi gave him a push: "Tell Georg about the article we have found." I was getting curious. It was an article Emi had discovered in a Japanese newspaper, the international edition of the Asahi Shinbun, and she had asked Masaaki: "Isn't this exactly the topic you and Georg are working on?" And indeed there was an announcement that a group in Tokyo had found superconductivity in lanthanum barium copper oxide.

At that moment, I got a bit nervous, on the one hand because I felt that a race had started. And we had not yet explored the obvious variations in the composition of our superconductors. We still wanted to check whether we really had the optimum dopant or whether strontium would do better than barium, but we were busy writing our paper on irreversibility. And our paper in which we showed that magnetic susceptibility confirms superconductivity was still pending.

On the other hand, this confirmation meant that there was no need to fight for acceptance of our discovery. I was also very pleased by the second news, which came from Paul Chu in Houston in early December. He sent a letter to Alex and me, and said that he was also convinced that this was superconductivity. He wrote that T_c could go even higher than 35 K. Experiments done under high pressure made him believe this.

10.2.11 Visiting the German Physical Society

The famous meeting of the American Physical Society in New York in March 1987 was called the Woodstock of Physics. At that time, the newspapers ran news on the latest findings in high-temperature superconductivity almost every day, and the

Physics communities were in a highly excited state. Especially since Paul Chu had announced his discovery of superconductivity in the yttrium barium copper oxide system beyond the temperature of the boiling point of liquid nitrogen.

I did not attend that meeting because I had to go to the German Physical Society meeting in Münster, to which I had submitted an abstract for a normal short presentation at the end of 1986. I was eagerly looking forward to this, since the University of Münster was my home university.

But in the meantime, because of the worldwide excitement that had started already in December, the organizing committee decided to have a special superconductivity session. They asked me whether I would be willing to give a review talk of half an hour to an hour. I agreed with pleasure.

During my stay in Münster, I took the opportunity to go home to see my family, and came back only a few minutes before the talk. When I came to the lecture hall, I found a crowd of excited people discussing in front of the door, they had no longer gotten access. The door was completely blocked. So I approached the first person and asked: "Excuse me, can you please let me pass?" He started laughing: "We are all waiting to get in for the superconductivity session, but there is no space any more." I replied: "Well I'm sorry, but I am the speaker in this session." And immediately, a corridor formed for me, and I could enter the lecture hall, which was completely packed, with the audience eagerly awaiting the special session. People who had not been able to get a seat were sitting on the stairs between seat rows and even on the floor, leaving almost no space for the speaker and the chairman. The speaker who had the talk before the special session was talking about Josephson junctions, and he must have been very surprised that he had such a big audience.

The session then almost developed into a science happening. The audience was excited and in a very good mood and followed my presentation with great attention and excitement as if I were talking about a miracle. People started shouting and applauding when I reported the latest news I had heard from my colleagues in the US only hours before—especially when I was talking about what happened when rumours spread after Paul Chu's discovery of the 90 K superconductor. Initially, the composition was secret, but "It's green" was the message that spread like wildfire through the science community. So everybody being involved in the search for new superconductors started to look for green compounds, and many mixed green chromium oxide into the new compounds. It turned out that the first samples by Chu and his group were multiphase. "It's black, too," was the message shortly after the pure 90 K superconductor had been identified, and the world was OK again. During the long discussion that followed my talk, also German colleagues from different research institutions presented their results, and at the end the organizer asked whether I could give a similar presentation including also these results the next day. A bigger auditorium, the biggest on the campus of my home university, should enable those who had not gotten access to the first session to get the first-hand information from one of the discoverers. This event, being in the spotlight so much, was a special experience. Yes, there is some truth in the statement made by observers that I came close to being like a rock star. And with the topic of high-temperature superconductivity getting lots of public attention through the press and television, many of the colleagues in the science community were like movie stars later on!

10.2.12 Nobel Prize

So many people were speculating about the Nobel Prize and told me that our work would qualify for this famous award. My reaction was always: "*You* can tell me anything you want, but *I* simply don't want to think of it." I found this the best way to protect myself. In this way I gave myself the chance to enjoy the success of our achievement, the appreciation and the recognition of the scientific community rather than becoming blocked by speculating and being disappointed year after year if someone else were to receive this honour.

So, when the Nobel Prize announcement was made, I could not believe that this was real! When the IBM chairman asked me shortly after the announcement had been made: "How do you feel?" I said: "Like in the clouds, as if disconnected from solid ground." The hours that followed were a great experience. I called Alex at a conference in Naples, Italy. I talked to my parents, who already had the press at home. My wife, who was working in the French part of Switzerland at that time, heard the news on the radio and immediately left her workplace. And when somebody asked her why she was leaving, she could only shout: "We got the Nobel Prize in Physics!" Two hours later, she showed up in the Lab in Rüschlikon, where we had champagne at a reception for all our colleagues and were waiting for Alex to be flown in from the conference in Naples, Italy. Then we spent the rest of the day, and the days that followed, answering questions from the media and giving interviews.

The Nobel Prize ceremony itself was quite an experience. But many things you do not..., what should I say, you cannot really experience fully because they engulf you and sweep you along. So I was very glad that in 1991 I was invited together with all other Nobel Laureates to celebrate the 90th anniversary of the Nobel Prize. It was also the year that Pierre-Gilles de Gennes was awarded the Nobel Prize in Physics. I could sit and watch from "behind the scene," really enjoy it all and relax as a spectator, and also fill in the gaps of what I had "missed" in 1987. In the days after the ceremony in 1987, I remember there was so much activity during the day and also at night. There were talks at universities, lunch and dinner invitations, and student parties at night. I sometimes did not sleep more than two hours or so. Normally one would come back late at night or early in the morning and get up at seven again. It was like in a dream. When I came back to Switzerland after two weeks, I was completely exhausted. I have been told that Alex was quite exhausted as well.

But also the time in 1987 before receiving the Prize was a relatively busy time. I never travelled that much later. Alex and I once counted our presentations. I myself came to 52 presentations—in nine months, all over the world.

10.2.13 Applications

Regarding the applications for the high-temperature superconductors, Alex and I were sharing the euphoric view of the majority of our colleagues in the new field.

10.2 His Own Story

We especially saw a great potential for reducing energy losses in the transmission of electricity. But when we started to talk about this in interviews, I recall that Alex Malozemoff, a colleague at the IBM Research Center in Yorktown Heights, gave us some advice: We should never emphasize the potential use for the transport of energy, as these superconducting ceramics would never be able to carry large currents because of their specific properties. He coined the term "Giant Flux Creep" to describe an effect that occurred because of the lack of strong pinning centres for magnetic flux lines, which made him pessimistic regarding the potential of these new materials for transporting electricity without loss.

Therefore, it was quite a surprise to me when one fine day I had him on the phone and he told me that he was leaving IBM to join a start-up company to develop high-temperature superconducting components and wires. It was his conviction that many of the applications people had been dreaming of would sooner or later become real. And indeed in the past two decades, the worldwide scientific effort has provided deep insight into the basic materials properties, which led to impressive progress in many segments of superconductor applications. Thin superconducting films took on an important role as model systems to study, among others, anisotropy effects in the pinning of magnetic flux lines and grain boundary effects. These key experiments had significant impact on the processing methods of bulk superconductors to enhance the critical current densities in wires.

Today there exist numerous test sites in many countries where superconducting cables are integrated into the power grid and reliably supply energy to tens of thousands of customers. I recently learned about a first power substation in China that is fully equipped with high-temperature superconducting components: Power transmission cables, fault current limiters, transformers and Superconducting Magnetic Energy Storage (SMES) devices have passed impressive performance tests, and the station is now serving three high-tech companies.

In addition to their impact for power transmission and management, wires in superconducting machinery, e.g., in persistent magnets, efficient generators and motors, will contribute to saving energy and using energy more efficiently and thus reduce environmental problems. The higher efficiency of a generator with superconducting wires is already boosting the output of a German hydropower plant.

Even more impressive is the gain in efficiency and productivity when in the metal industry a superconducting magnet is used to create eddy currents in a rotating metal bar which provides fast and homogeneous heating. This example shows that there is still room for the development of creative new applications.

High-temperature superconductors, because of their unique properties, high critical current density in combination with low thermal conductivity, have also made their way into the LHC facility at CERN. Here they serve as current leads to bridge the difference between room temperature and 2 K and feed currents of up to 13,000 ampere to the 1200 helium-cooled bending magnets.

And last but not least, there are numerous applications which are based on thin films in electronic devices and sensors, and work in the low-current regime. Already years ago, I saw compact test units in China that were equipped with closed-cycle refrigerators to cool high-temperature superconducting films and which served as filters in base stations for wireless communications. And already in 1993, I was impressed by a SQUID setup I saw in Jülich, Germany. I had my heart inspected by this device which recorded an MCG (magnetocardiogram), and asked my medical doctor for a copy of my ECG (electrocardiogram) as well. And even as a non-specialist, one could see that the magnetically recorded signal contained much more information with an even higher resolution in time than a traditional ECG does. Nobody knows what this additional information means, but the information is there, and so we have to find out and learn how to read and interpret it.

High-temperature superconductivity has come a long way from science and laboratory experiments to prototypes and products. After more than two decades of research and development, the field offers a fascinating spectrum of applications—we just need to embrace them. Then I am sure that superconducting technology will grow to be a key technology of the 21st century. I am happy and also a bit proud that through our discovery in Rüschlikon, we have provided some of the basis for this exciting development.

Chapter 11
K. Alexander Müller: Discovery of Cuprate Superconductors

"I heard this talk about Jahn-Teller polarons by Harry Thomas in Erice"

Fig. 11.1 K. Alex Müller. In front of diplomas for honorary doctorates from 22 universities

11.1 Biographical Notes

K. Alexander (Alex) Müller shared the Nobel Prize in Physics for 1987 with J. Georg Bednorz *"for their important break-through in the discovery of superconductivity in ceramic materials."* See also Chap. 10.

Müller was born in Basel, Switzerland in 1927, and lived first in Salzburg where his father studied music, later in Lugano, where he became fluent in the Italian language. His mother died when he was eleven, after which he attended Evangelical College in Schiers, in the Swiss mountains. He remained there until the end of the war. He was fascinated by the radio, and wanted to become an electrical engineer, but his chemistry tutor, Dr Saurer, convinced him to study physics. After military service he enrolled in the Physics and Mathematics Department of the Swiss Federal Institute of Technology (ETH). The freshman class was three times too big, and the process of elimination was correspondingly tough. They were called the "atom bomb semester" for obvious reasons. Müller had excellent teachers, like Scherrer, Känzig and Pauli, and did his diploma work with Professor G. Busch on the Hall Effect in grey tin, followed later by PhD work on paramagnetic resonance (EPR) in Busch's group. Here he identified an impurity present in the perovskite $SrTiO_3$, a fact he took much advantage of later. Upon completion of his PhD and after graduation in 1958 he worked at Battelle Memorial Institute in Geneva, whereafter he came to the IBM laboratory in Rüschlikon in 1963. He remained there until his official retirement from IBM, after which he continued as a professor at University of Zurich.

Alex Müller was a key person in the research which took place in the late 1960s and in the 1970s and early 1980s on understanding the critical properties of phase transitions in solids. Again $SrTiO_3$ was the vehicle, and it became the best studied of all, especially its properties related to the structural phase transition near 105 K. With Harry Thomas he identified the order parameter and worked out the Landau theory for this system. He was and is a world leading scientist as far as structural transitions in perovskites is concerned. This competence was not wasted, as it turned out, when he undertook the challenge with Bednorz to find superconductivity in oxides. From the time superconductivity was discovered in oxygen deficient $SrTiO_3$ at Bell Labs in 1964, he had an eye on this subject, but did not get directly involved in superconductivity until his 2-year long sabbatical at the IBM lab in Yorktown Heights at the end of the 1970s, at which time he and a group of colleagues studied Tinkham's textbook from A to Z, as he said to us: "... like a graduate student after the age of 50." Now he started research on superconductivity for the first time, in granular aluminium.

His interest in the subject did not diminish after this. Some time after his return to the IBM lab in Rüschlikon, having heard a talk by Harry Thomas at a meeting in Erice, he was inspired to invite Georg Bednorz to collaborate in a search for superconductors among Jahn-Teller perovskites. We refer to his own account, as well as that by Bednorz in the present book about the ensuing progress. The work that Binnig and Bednorz had done on his "old favourite" among perovskites, $SrTiO_3$— a work he had followed closely as a manager—was also on his mind when he suggested the collaboration which would turn out such spectacular results, ending with the sensational developments in late 1986 and in early 1987: The discovery of record breaking high-T_c perovskite superconductivity in $La_{2-x}Ba_xCuO_4$.

Alex Müller has achieved the rare position to be a world-leading scientist in two different fields of condensed matter physics. Those who have had the privilege to know him, have experienced his profound ability to combine knowledge from different areas of physics into a penetrating understanding of complicated subjects. The award of the Nobel Prize in physics to Bednorz and Müller in 1987, attests to the fact that the spectacularly important and unexpected is often to be found in such combination of knowledge. Alex Müller holds on to his original ideas about the (bi)polaronic mechanism for superconductivity in the cuprate superconductors, a view that undeniably led to their great success. In his view, the observed isotope effect, as well as the so-called stripe domains attest to the correctness of this basis for superconductivity in cuprate superconductors. His latest paper on the subject was published online in J. Supercond Nov Magn in March, at the age of 84. Müller is the recipient of numerous awards in addition to the Nobel Prize. He holds honorary doctor degrees at 22 universities.

11.2 His Own Story

11.2.1 Background

The background of my family is that both on mother's and father's side they were businessmen essentially. And I am kind of a delta function in my family with regard to science. I was living with my mother in Lugano, Switzerland after the separation of my father and my mother, and she passed away when I was eleven years old. At that time my father made a very important decision to send me to a college in the Alps. From the first grade to the baccalaureate to the matura, I was there.

At the age of 15 I was interested in radio, an enthusiast. I had built already, while my mother was still living, at the age of nine, the first receiver, a one tube receiver. On and off I continued that interest, and I wanted to become a radio engineer.

Three months before final examination, my chemistry teacher asked me: "What are you going to do?" And I said, "I have an interest to become a radio engineer." And he said, "why?" I said, "I'm interested to know what happens in these radio tubes and in these condensers." And then he said, "Well, if you're interested in that you better study physics, because then you will understand more of the electronics." And he looked at me and said, "You are sufficiently good in mathematics to go through the physics." And this is what I did.

11.2.2 At ETH

I inscribed myself at ETH. Normally there were about a dozen students who studied physics, and my family had no idea what this meant. They only said, "Maybe you become a high school teacher in physics." That's all they knew. In the meantime the Americans had blown their atomic bombs in Japan, and so there was a huge interest in nuclear physics, and in atomic physics, which was my interest. So instead of being a dozen, we were four dozen students. We were called the "atom bomb semester."

This was in '46, just after this had happened. Of course the mathematics teachers at the physics department were not prepared for such a large number of students. So they had internally decided to be really tough on us, making the study difficult—not wanting too many students—to reduce the number. I studied condensed matter physics, being interested in atomic physics, electronics and so on, *Festkörper Physik* called. I got into that, perhaps after the first two years, because the lectures of my later PhD professor Busch, were very good, very clear, whereas the nuclear lectures were not so interesting, and also it wasn't my primary interest.

Bush was already well established in solid state physics. He had originally done ferroelectrics and was the co-discoverer with Scherrer of ferroelectricity in potassium di-hydrogen phosphate, KDP, in 1935. Then they had a separation, and Bush decided to do semiconductor physics where he was the discoverer of impurity band conduction in silicon carbide. This was how I entered that field.

11.2.3 $SrTiO_3$ and Superconductivity Preliminaries

I should point out that superconductivity was not in the curriculum at all. At our institute superconductivity came from Oxford with Jan Olsen, a Danish physicist. My only connection with him was during my PhD studies when I did electron spin resonance (ESR). He had installed a Collins liquefier, and I wanted to do ESR at helium temperature as the first one in Switzerland. So Olsen helped me transfer helium. That was all. He wrote a little textbook too, and he had a large number of successful students. Heini Rohrer was one of them.

I did my PhD on electron spin resonance in strontium titanate, $SrTiO_3$. My thesis was finished in '57, and published in '58. A bit later on, at Bell Labs, superconductivity was discovered in strontium titanate, after they reduced it. This was in '64. It aroused my interest a bit, and I went to Olsen and said, "Hey look, is it possible with superconductivity in $SrTiO_3$?" He was very sceptical. He said, "Probably they have used so much titanium that there are some titanium filaments in it." Since then I followed a bit what happened.

The reduced $SrTiO_3$ is oxygen deficient, with a T_c of 0.3 K, and later Georg Bednorz with Gerd Binnig and also Alexis Baratoff, all working here at the IBM lab, picked up the problem. I was not involved, but I was the manager of the physics department in the IBM laboratory in Rüschlikon at the time. Georg made single crystals, niobium doped, and with that T_c went up to 1 Kelvin, which was enormous, with a bare 10^{19} per cubic centimetre concentration of carriers. And then, of course, they were thinking of going to much higher T_c by doping higher. What happened was, they doped further and discovered two-band superconductivity. And if you doped even higher, T_c went to zero. Now one understood it, because the electron-phonon coupling, which is in principle classical, was huge, much stronger than in the metals, and with 10^{19} carriers per cubic centimetre, you can't get superconductivity. Later, the reason was found, that because of this low doping, the plasma edge was below the highest phonons, and then the highest phonons made the coupling. When they doped higher, the plasmon passed the highest phonons, and you get the classical shielding that you find in every book, and the T_c disappeared.

This was kind of disappointing, but at least Georg retained his hopes even if it did not work out so far. And then in '83, when I heard this talk about Jahn-Teller polarons by Harry Thomas at a meeting in Erice, I said, maybe one can do something.

Then, when I came back, I said to Georg, "Should we not try something with Jahn Teller ions?" And within two hours he said yes. He was interested. But the basics were the background, you see.

And then came this concept of Jahn-Teller polarons, and we started work for two or three years, but under the table; we didn't say it to anybody. Georg was a young man, and you know how things are in such labs. You don't want to hurt a young man's carrier.

My interest since the discovery of superconductivity in $SrTiO_3$ in '64 was of course there. By that time I had become a sort of an expert in strontium titanate. And so very early—it must have been the beginning of the seventies—I went to Oslo, to Jens Feder. He had set up some crystallographic lab, and we said, "Let's try lanthanum titanate; let the lanthanum dope electrons into the system." He tried a bit, but not much, so it went to sleep. Nowadays I understand the results Georg obtained. He has investigated the whole phase of titanate doping in lanthanum titanate, going from $O_{2.5}$ to O_3. And it doesn't go. But now I understand why, it's because the titanate doesn't form bi-polarons, but this is what I know now, since a year, anyway.

11.2.4 Yorktown Heights: Learning by Doing. Conventional Superconductivity

Later, I spent two years at the IBM Thomas J. Watson Lab in Yorktown Heights in New York, where I worked on conventional superconductors. I was over fifty and I had done a lot of work in all fields of condensed matter, but never in superconductivity. And the reason is, I didn't understand it.

Just before leaving for the United States I had bought a book by Werner Buckel, which is a nice introduction. And then we thought we could do some ESR at the transition from normal to superconducting state. But then you need a metal, which shows electron spin resonance, and it must be a light metal, because otherwise the spin-orbit coupling is too strong. In the heavier metals you can't see it. In order to be able to observe something, I thought of granular aluminium as a superconductor, which is a very interesting substance. Also, before going to the States I had visited the lab of Guy Deutcher in Tel Aviv, and he and a colleague had done really nice work on granular aluminium, so he gave me some samples. Due to the reactivity of pure aluminium, when the powder is made, each grain is immediately covered with aluminium oxide. So I had a granular substance with a film of oxide. And between the film and the grain you have to establish coherence, and the smaller the grain is, the higher the T_c is, that is the funny thing. You can go from 1 to 7 Kelvin.

We put it in the cavity and we didn't see any ESR, the relaxation time was too short. But we found the microwave response of this granular aluminium, and we could measure at microwave frequencies, both the impedance and the resistivity of the substance. It was done with Mel Pomeranz, and with Alex Malozemoff who also had no idea about superconductivity at that time. So I proposed to form a group to understand superconductivity. We were four, Mel Pomeranz, Alex Malozemoff, myself, and Elena Alessandrini.

And so we said, "Well, let's take Michael Tinkham's book and start at page one. We'll meet twice a week and go through, page by page." And this is what we did. So all followed superconductivity in this way, because often when we were doing experiments Mel and I had not the slightest idea what we had observed. So it was like, after fifty, I was a PhD student. We started from fresh. Finally, we concluded, by measuring these impedances and so on, that it was a percolative superconductor. And this was at variance with what the opinion was at the time, that it was a question of phase coherence coming in. And we said; no it's just percolative patches.

However, the management felt that we should not submit that. They had hired Paul Horn, now a senior vice president of research at IBM, who was the big shot regarding phase coherence also in superconductivity, and he didn't believe it. Eventually we submitted, and we got it accepted within three weeks. One of the referees was quite enthusiastic about it, and this was the first time I got an invitation for a talk at Harvard, because one of the referees was Michael Thinkam!

This was my start in superconductivity. And then I came here to the University of Zurich afterwards, and I lectured superconductivity for one semester. I had become interested in the field.

11.2.5 *The New Beginning: Jahn-Teller Polarons*

It was after this I went to a meeting in Erice, Sicily and heard Harry Thomas give a talk on polarons. He and Höck and Nicksch had proposed a Jahn-Teller polaron theory, and at that time I told him: "Look, I understand the oxides and I'm sure there are ions which show the Jahn-Teller effect, but whether they are mobile I don't know." I tried to prove that polarons existed in these materials, which by now in manganites everybody is convinced there are. But, at the time it was not so.

When Georg and I had written the first superconductivity paper, we gave it to our colleague, Eric Courtens, who then gave it to Harry Thomas to read, and he didn't believe so much in it for sound reasons, because they had computed the effective mass with variational methods, and it turned out to be very large. So Harry said it is probably not possible. And, of course, now we understand that what is much more mobile is the bipolaron, it makes such a creepy spider-like motion. And, of course, if the mobility is larger, you have more complications, you have fermions at the same time.

I had worked a lot in EPR on Ni^{3+} and it was a strong Jahn-Teller ion. So at the beginning with Georg Bednorz, we tried to use Ni^{3+}. Now, what was known was that $LaNiO_3$ is a perovskite, and it's a metal. However, because of the overlap between nickel and oxygen orbitals, you have a very large band. And we have known in the Jahn-Teller business for a long time, that if the band is broad, the Jahn-Teller effect is quenched. And there is no superconductivity.

Then we tried with Georg to dilute it, instead of having lanthanum nickel oxide, we replaced nickel with some aluminium, to narrow the band. More or less that is what Bertram Battlogg has now done with the fullerenes. Fullerenes are like balls, which can be pulled apart by having some chemical in between, and he gets a thinner band, and therefore it is more polaronic.

And so we tried that. However, whereas in the fullerenes you can do it regularly, where you have something in between, in our case it was stochastic where the aluminium was, and therefore the potential could vary in a stochastic way, and it could localize. So basically you are on the other side of the Anderson transition, where it localizes. We tried for almost two years on nickelates.

For one thing it was doping on the wrong place in the lattice. That is one thing, and the other thing was the three dimensional structure, whereas the ones we know now are layered. So there are two things that hindered it. And nickelates, which are two-dimensional with Ni^{3+}, have still not been found, but one may be able to find them.

11.2.6 The Discovery

At this time I was not yet aware of Ginzburg's work in this area. I learned about that in '87. In the summer there was a Trieste meeting. Ginzburg was there, and he told us that already in some early work he had recommended two-dimensional or layered structures as a way to reach high T_c.

We got there even so. This was—probably Georg has mentioned that—by chance this mixture of compounds, which originally were thought to be catalysts, because the oxygen mobility is very large. He now found the material from the Raveau group, and immediately understood that this was the way to attack the problem. It was essential.

The Raveau group later tried out their samples, and also people at Thompson CSFF in Paris found that 75 % of their samples were superconducting!

They had not had any idea. They were thinking about oxygen mobility of iron, you know. They were not looking, they were not interested in superconductivity at all at that time, so one can not blame them for not finding it at all. And they had a nice career later, it was a nice institute.

So, eventually Georg started finding conductivity anomalies in these compounds. It looked very interesting, but we had no susceptometer yet, therefore we could not definitely confirm superconductivity. So, we decided to submit a publication on just the resistivity results. The journal was Zeitschrift für Physik, and if you submit a normal article you get galley proofs after four months. The delivery time for a susceptometer would be only three months, so we knew by the time the proofs would arrive we would have this thing either confirmed or disconfirmed, and if it was not correct, we could dump the whole thing. And so it went.

And of course I was old enough to know how the game is, the referees etc, and to shield oneself from that. Basically, those who knew it were Georg, who had done the experiments, and myself, and Eric Courtens, as a manager at the lab, and Harry Thomas, as a referee, who had invented the polaron. All of them were IBM related people who would be loyal.

11.2.7 Definite Proof, Growing Attention

For definite proof of superconductivity, susceptibility is the ultimate test. This was carried out about a month earlier, before the Nobel Prize award to our colleagues Rohrer and Binnig was announced. At that time I had already stepped back, having been head of the physics department, but I wanted to do these things. Heinrich Rohrer had succeeded me as head of the department. And so we wrote up the susceptibility results, and I think the day that he got the telephone call from Stockholm, Georg went up to him with the papers to get his signature so we could publish it, with the susceptibility. But then the Europhysics people did a very bad job in publishing the next article, with the consequence that the Japanese, with Tanaka, who had also confirmed it, appeared before ours. Our submission date was a month or so earlier, October 22. And that was sufficient.

11.2.8 Recognition and Priorities

But it was not big news until there appeared a note in a Japanese newspaper. We did not know. But we had a Japanese guy who was working with us, Takashige, with whom we later published the susceptibility of the irreversible behaviour in the substance, had a forwarded issue of this newspaper. It was a large newspaper, and one morning he came and showed it to us. I think that he then called Tanaka, and Tanaka was at first a bit resistant. It was in our favour, that when they submitted their paper, it was six months after we had this original thing. So, in this respect there was no big discussion.

Then in parallel, also Paul Chu in Houston had learned about it and did some work on the lanthanum compound, applying chemical pressure to increase T_c. With the higher pressure he had gotten T_c up from 30 to 50 K, and he called me up, and said: "That's great, with that we will go higher."

Georg also had wanted to do it, but by then we were already overflowed with invitations for talks and so on. And so it was Houston and also others who did that sort of compound. The pressure on Georg and me was already quite substantial, and became enormous.

Before that, I had an invitation to Köln, which I had gotten half a year earlier, and the host was interested in granular aluminium. So, I should have talked about granular aluminium. But then, after the submission, I change my subject to this. And the effect was that we were in a relatively large lecture hall and less than a dozen people came. And so I explained: "Look, these are the resistivity measurement, and this is the susceptibility, it looks to us like superconductivity. If somebody feels differently," I said to the people there—Mühlschlegel was there, I think—"he should say so." But nobody said anything. And then the chairman said, "thanks for this interesting suggestion," and that was it.

And a week later Georg gave a talk in Berlin, with the same effect, that there came a few people. If these people had been not just sceptical they could have tried themselves, and Europe would have had quite an advantage, because this was before the United States woke up. This was in the fall of '86. After this it was quite a thing. Within a year after I got the Nobel Prize together with Georg, I had received 850 congratulations. I don't remember how many he got, he also got a lot. Of course we printed some cards saying thank you, but half of the people I knew, so I had to write something. So, it was quite an involvement. Staying ahead was a challenge under these circumstances.

Another important discovery was the irreversibility line. I was interested in statistical mechanics, and therefore all these notions were familiar to me, and also some theories. We wrote a paper in Physical Review Letters on this, and it started a whole other branch in the field. It turned out to be really important. The challenge was to control this behaviour.

Here at the university, Hugo Keller, with the muon rotation, has proved a lot in this field. They found—they were the first, although it is not so noted in the States— but they were the first to find that the vortex lattice melts on heating. It is first order, and the muon rotation showed that first. So, Hugo did quite a bit in this area, which I don't follow anymore.

In the Nobel lecture I explained the concept which led us to this. And specially, the two years that followed were very comforting in this respect. What was not seen, neither by the theoretician Harry Thomas, nor us, is that one of these Jahn-Teller polarons first meets another one and creates a bipolaron, which is relatively evident. Somebody else at the time, we felt, who could find this high-T_c superconductivity, was John Goodenough, that there are elastic interactions between the two, because there are ions that are displaced. Therefore you know the energy by binding two together. Which at very low dopings is in the order of an electron volt, and if you dope higher, of course the interactions surrounding these gets smaller.

But this is now quite documented, because vibronic means that the electric energy and the vibration energy should be similar. And now there are data from both sides: the elastic neutron scattering, which probes the phononic part, where you see a clear anomaly, and at the same time also with photoemission, by Lanzara in Chen's group at Standford.

11.2.9 *The Future, as Seen in 2001*

For six years, till two years ago, I was a consultant with American Superconductors, and I followed that quite closely, I would say. For the high power application the current carrying capacity of these cables and wires goes up linearly with time. This is called Malozemoff's law, named after the technical director of the company, Alex Malozemoff, who was previously at the IBM lab in New York. And that is nice, as long as you have to coat it with silver. You have to use metal mantel that let oxygen diffuse through it, and this is the only metal you can use, especially in those copper oxides, with bismuth oxide you have to adjust the oxygen inside very accurately. This is a difficulty because silver is expensive, and it pushes up the number of dollars per ampere you carry through these wires.

There are now alternative techniques of depositing the superconductor on substrates, but these techniques are not yet so far developed that you can use it commercially. The first one I mentioned you can use, but in order to keep the number of amperes per dollar high, you need still much higher capability, and therefore I would say to have an alternative compound, where you do not have to adjust the oxygen that accurately, that you can make these powder-in-tube things, without at the end having this very delicate annealing procedures in oxygen, is something which is necessary.

Also, because the way they are made now, is not with one filament, but many filaments, and between these filaments there is silver, and it causes AC losses, and in order to have AC cables, this is clearly a difficulty. To improve that, you need to shield these filaments, as has only been partially successful until now; or to find a compound which is not so sensitive to oxygen content, is relevant.

Of course, what they have now, in part, works. Let's say in Detroit they have now a few hundred yards long power line which works. But if you would like to use it in AC applications—I think this became clear at the meeting in Norway a few years ago—this is a difficulty in such applications.

And I think the situation remains the same, there are now applications, let's say for current leads to low temperature conventional superconductors, which are used quite successfully, for instance in systems were you store energy magnetically, in MRI type applications, where the current leads are high-T_c compounds, so you don't have a big heat leak. And I think this sector for American Superconductors is commercially profitable, but it is of course only a segment of what they do. I think at least this MRI thing is certainly going to stay, and the current applications probably too.

Another thing they do, is chemical deposition, but the production is relatively slow, one hundred kilometres per year or so. Another difficulty is that this is a replacement technology for the present high current applications, and you are probably not going to throw out a conventional transformer, which worked for fifty years. In Siemens they are looking forward to having transformers on locomotives, were they gain in weight by a factor two or three, and even with the present technology they say it is interesting. Their main difficulty is the discussion with management of railways, because when these people learn that on the inside there is a little bit nitrogen, they become sceptical. So they have to mantel it so they don't see the nitrogen, despite that these transformers are advantageous from several points of view, according to Siemens. It's not dangerous like a normal transformer, which can burn the oil. And it's lighter and also to detain it is ok.

11.2.10 Concluding

Now, to terminate our discussion, I will quote Weisskopf. He was a great man, and the last book he published has the title "On the privilege of being a physicist." And he has therein a dozen of talks he has given on various subjects, and one of them is

11.2 His Own Story

Fig. 11.2 Alex Müller visiting the author in Trondheim, in December 1987 just after receiving the Nobel Prize in Stockholm. Müller *left*, the author to the *right*

on technical innovation. There he says: It takes on the average ten to twenty years before a discovery becomes really working. And he gave some examples. And I can add some other ones. For instance, one of the last Nobel Prizes was on the very large scale integration. It took about twenty years from somebody conceived it till it worked on a useful scale. The transistor was the same thing, from the transistor was demonstrated till it was competitive with good radio tubes took well over ten years. After ten years they were able to purify silicon so much that you could start using it commercially.

And the first transistors you could not use at high frequency, etc. And it goes back further, even to the steam engine. Watts took twenty years to put it in use. And the average of twenty years is a normal one. And there are extremes the other way. One extreme, the shortest I have found, is X-rays. With X-rays, broken bones were photographed within half a year after Röntgen found the effect. And the longest is fuel cells. Fuel cells were invented hundred and fifty years ago and are still not commercial. So there is a huge scatter, but on the average twenty years is reasonable.

And to me it looks like that this average may be met by the high temperature superconductors.

Chapter 12
The Anderson-Higgs Mechanism for the Meissner Effect in Superconductors

A. Sudbø

In this day and age, when the world recently has celebrated what is most probably the detection of the predicted Higgs-boson of the Standard Model in the Large-Hadron Collider at CERN, it is only fit to mention the Higgs-boson and the Higgs-mechanism in a popular text on superconductivity. The reason is that the Higgs-mechanism plays out in a spectacular manner in the remarkable electrodynamics we find in superconductors, most notably type-II superconductors. Moreover, superconductivity is the first known physical phenomenon (albeit non-relativistic) which is a direct manifestation of what has become known as the Anderson-Higgs mechanism, and which was understood as such.

Although this is not a technical book, it is beneficial for the following brief comment to inject a slight amount of mathematics into the narrative. A very useful mathematical formulation in this context is the famous phenomenological theory of superconductors written down by Lev D. Landau and Vitaly L. Ginzburg in 1950. It is based solely on symmetry principles with no knowledge of any microscopic theory of superconductivity. It is the simplest possible description of a quantum mechanical charged many-body system which may condense into a macroscopic phase-coherent ground state. The Ginzburg-Landau model is defined by the Lagrangian

$$\mathcal{L} = \frac{|\mathbf{D}\Psi(\mathbf{r})|^2}{2M} + V\big(\{\Psi(\mathbf{r})\}\big) + \frac{1}{2}(\nabla \times \mathbf{A})^2, \tag{12.1}$$

where $\Psi(\mathbf{r})$ is the complex scalar field associated with the condensate, M is the mass of the condensate, and $\mathbf{D} = \nabla - ie\mathbf{A}$. Moreover, $V(\{\Psi(\mathbf{r})\})$ is the potential term. The model has a *local* $U(1)$-symmetry, since it is invariant under the transformation $\Psi \to \Psi e^{i\phi(\mathbf{r})}$, $\mathbf{A} \to \mathbf{A} + \nabla \phi/e$.

This means that the order parameter that works for a superfluid, namely the global phase-stiffness of the system, will always be zero in the charged case, i.e.

A. Sudbø (✉)
Department of Physics, Norwegian University of Science and Technology, 7491 Trondheim, Norway
e-mail: asle.sudbo@phys.ntnu.no

a superconductor. This is so, since any phase-twist in Ψ can be compensated with impunity by an adjustment of the vector potential $\mathbf{A}$. Instead, another way of measuring order in a superconductor comes into play, in the following way. Expanding the kinetic energy, we obtain a term $e^2|\Psi|^2\mathbf{A}^2/2M$. This term is quadratic in the vector potential, and has precisely the standard form for a mass-term $m_A^2\mathbf{A}^2/2$ for $\mathbf{A}$, where the mass m_A of the vector potential is given by $m_A^2 = e^2\langle|\Psi|^2\rangle/M$, where $\langle\ldots\rangle$ denotes a statistical average. This mass is generated by the expectation value of the condensate density $\langle|\Psi|^2\rangle$. In a mean-field theory, the onset of this expectation value marks the onset of superconductivity. The appearance of such a mass-term breaks the local $U(1)$-symmetry above.

Hence, at the mean-field level, we see that the onset of superconductivity generates a mass for the vector potential, such that "the photon becomes massive"! This mass represents the inverse length scale over which a magnetic field is able to penetrate into the superconductor, and is to be identified with the London penetration length λ_L, i.e. $\lambda_L = m_A^{-1}$. The dynamical generation of the mass of the vector potential through the onset of a finite condensate density is therefore nothing but the Meissner-effect.

We may bring out the role of the all-important *phase* of the condensate field Ψ in the following manner. Let us rewrite Ψ as follows, namely $\Psi = \rho e^{i\theta}$. Setting ρ equal to a constant everywhere in space and determined by the potential minimum in Eq. 12.1, this equation may be written as (omitting constant terms)

$$\mathcal{L} = \frac{\rho^2}{2M}(\nabla\theta - e\mathbf{A})^2 + \frac{1}{2}(\nabla \times \mathbf{A})^2. \tag{12.2}$$

Now, setting $e\mathbf{A}' = e\mathbf{A} - \nabla\theta$ (such that $\nabla \times \mathbf{A} = \nabla \times \mathbf{A}'$), we see that the Goldstone mode θ has vanished from the action, having been subsumed in the gauge-field, while the gauge field has become massive with a mass-term given by $(e^2\rho^2/2M)(\mathbf{A}')^2$. This is identical to the mass-term given previously. (The above assumes that the phase fluctuations $\nabla\theta$ subsumed in $\mathbf{A}$ are curl-free.)

This way of extracting a mass for the photon at the mean-field level is identical in spirit to the way of extracting masses for the $W^\pm$- and Z-bosons in the Standard Model through the expectation value of the scalar Higgs-field. The energy scale at which fluctuations in the Higgs-field of the Standard Model become important is many orders of magnitude beyond what is attainable in any particle accelerator, so a treatment of the masses generated by the Higgs-field at mean-field is entirely adequate.

In superconductors, on the other hand, the situation is in some sense much more interesting. This is particularly so in extreme type-II superconductors with high critical temperatures, of which the famous cuprates are the prime examples. In such systems, critical fluctuations are prominent in the Cooper-pair wave function Ψ, which is the Higgs-field of the superconductor. While in high-energy particle physics the Higgs-field represents some omnipresent condensate permeating the entire universe up to astronomical energy scales, the "universe" inside a superconductor has a tunable Higgs-condensate which can be made to vanish at the "symmetry-restoring" critical temperature (where the mass of the photon vanishes, equivalently the Meissner-effect disappears) which is easily attainable in a terrestrial table-top experiment. In type-II superconductors, most notably high-T_c extreme type-II superconductors, the Higgs-generated mass of the photon is driven to zero via the proliferation of a particular and interesting type of topological defects in the superconducting pairing field Ψ, namely closed vortex loops. These vortex-loops have much in common with cosmological strings.

Thus, the study of "Symmetry-restoration" in extreme type-II superconductors as the system is heated from low temperatures and up through the critical temperature, may in fact shed some light on the early Universe right after the Big Bang, where temperatures were so high that the cosmological Higgs-condensate had not yet been established through spontaneous symmetry breaking. The Ginzburg-Landau theory serves as a useful "toy-model" for such studies.

Much can perhaps be discussed about the origins of the Higgs-mechanism. The lack of any notion of Goldstone bosons in the 1950-paper by Landau and Ginzburg notwithstanding, it is difficult to imagine that these authors would have failed to notice how a mass-term for the photon is being generated by an expectation value of the condensate field Ψ, and the implications this has for the Meissner-effect. It is but the simplest example of a mass being generated via a condensate field, which may serve as a quite general definition of the Higgs-mechanism. Confronting the Goldstone theorem,[1] as well as extending the mechanism to fully relativistic theories would of course have to await the seminal works of Anderson (1963), Higgs (1964), as well as Guralnik, Hagen and Kibble (1964). The principle of the mechanism, namely mass-generation for gauge-bosons due to the presence of a scalar condensate, does not at all depend on whether the system is relativistic or non-relativistic, though.

Another point to note is that the Standard Model of particle physics is a purely phenomenological model, with the same status that the Ginzburg-Landau theory had prior to the BCS[2] theory of superconductivity. The BCS theory provided a solid microscopic foundation for the Ginzburg-Landau theory. No corresponding microscopic foundation is available for the Standard Model. People working on superconductivity should therefore, perhaps more than anyone else, feel a deep ownership and kinship to the Anderson-Higgs mechanism.

---

[1] Goldstones theorem states that whenever a continuous symmetry is spontaneously broken, massless particles will appear in the spectrum of the system. Well-known examples are phonons when a liquid freezes to a crystal, and magnons when a paramagnet becomes ferromagnetic/antiferromagnetic. In superfluids, Goldstone modes appear as phase-fluctuations of the superfluid order parameter when superfluidity sets in, and are acoustical phonons. In the presence of gauge fields, like in the case of superconductors, the Goldstone modes are subsumed on the gauge field (Eq. 12.2 above) and do not appear in the spectrum.

[2] Named for J. Bardeen, L. Cooper, and J. R. Schrieffer, who came up with a microscopic theory of superconductivity in 1957. This theory was used by L. P. Gorkov to derive the Ginzburg-Landau theory from microcopics shortly thereafter.

Chapter 13
Concluding Remarks

As indicated in the Preface, this book was intended to cover aspects of both science and science personalities, or as the title says: *Discoveries and discoverers*. Naturally, each personal story is entirely different from the next. Also, we have to bear in mind the conditions under which the scientists grew up. For most of them this was before and during World War 2. Technology was an inspiration for several of them. The most advanced technologies developed during that period were the radio, radar and the nuclear bomb. Some were motivated to study electrical engineering through the prospect of communicating with the world far away. Most likely the future possibilities promised by the invention of the transistor also played an important role. Reaching out globally by radio waves was a fascinating promise. None of them mentioned nuclear physics as a motivation for their original interest in science. Yet, one of them seemed to have wanted to work on the bomb if allowed, another was a co-inventor of the first fully "successful" hydrogen bomb, a fact he did not mention in the interview since we were discussing superconductivity.

The most impressive and touching human aspects of the stories we have been told, are about tragic or unfortunate childhood experiences, like the loss of mother at a very young age, and lack of adequate schooling up to almost 12 years of age. Three of the ten laureates had such experiences. It is a great tribute to the resilience, endurance and adaptability of the human mind that such adversities could be overcome to such an extreme extent as to reach the world's highest level of scientific achievement. One of them built his first radio at the age of nine. Even in our time this offers a strong signal of encouragements and hope to youths who have experienced severe loss, inadequate schooling or other obstacles in childhood. Dreams may still be fulfilled. And youth may have its advantage: In three cases the breakthroughs in this demanding physics discipline were achieved by students, one of them a mechanical engineer just beginning to study physics.

No doubt, all ten laureates were blessed with unusual talent, and several were endowed with a great ability to endure hardship. Towards old age, those still among us have remained active all the way, in their 70's, in their 80's and even into their 90's; for some under conditions of poor health. One of them, at the age of 83, recently reported to me from the US: "I drive 32 km to work every day, if I am not travelling."

About a year ago he gave lectures in South Korea. Phil Anderson, 89, recently supplied detailed comments for this book on his own work in the late 1950s and early 1960s, related to symmetry breaking and the Higgs mechanism. Vitaly Ginzburg gave a full conference talk, although ill at the time, from his home in his 93rd year, the same year he died. Alex Müller, near age 85, recently was very pleased to hand me his latest scientific publication—his last he said—in which he presents further evidence in support of his original ideas about the cuprate high-T_c superconductivity mechanism. One might ask what message such endurance carries in view the widespread demand for early retirement in some of our modern, affluent societies with quickly increasing lifetime expectancy.

A journalist who recently interviewed me about Nobel laureate Ivar Giaever for a TV program, was very keen on finding whether Giaever's work is useful in an easily recognisable manner. I told him that the main point of the award had been to honour Giaever for offering conclusive confirmation of the BCS-theory, not for the possible practical consequences of his discovery. So, was his work useful?

Quite naturally, scientists and the general public will hold somewhat different views on what is useful. A natural scientist's job is to unravel truths about how Nature works. In his and her mind that *is* useful, whether or not the results are practically applicable, since it fills in the white patches in the master painting of the Universe. In contrast, we tend to think that usefulness in the eye of the general public is synonymous with applicability in devices, instruments and machines. But *fascination* with surprising discoveries and achievements also ranks high in the minds of most people. Fascinating facts and phenomena at the outskirts of our imagination meet a need to look for our place in the Universe, and are in this sense "useful." The best proof of this is the fact that the public is willing to support amazing science, like the Hubbard space telescope looking far out and back in time, and the Large Hadron Collider looking deep into matter and far back towards The Big Bang, history's most extravagant human scientific endeavours.

Science is first and foremost about ideas. New ideas arise sometimes from experiment sometimes from theory. Physics usually progresses in alternating steps of observation and theory. An experimental discovery like superconductivity was at first just a new and very surprising fact, then a challenge requiring new theory, and finally raised a demand on experimentalists to provide confirmation of theory. But in addition to an explanation of the basic mechanism, a plethora of successive discoveries followed, almost as exciting and surprising as the existential superconductivity mechanism itself. Let us be reminded of some of these discoveries: Phenomenological theories applicable to infinitely large quantum systems by the London brothers, and in particular by Ginzburg and Landau, Cooper pairs, flux quantization, the Abrikosov lattice with magnetic phase diagrams, flux pinning, quasi particle tunnelling, Cooper pair tunnelling, the physical reality and consequences of quantum mechanical phase, the applicability of Ginzburg-Landau type theory in high energy physics all the way back to the Big Bang and the Higgs mechanism, the applicability of BCS-theory in nuclear matter and in neutron stars, as well as supporting the prospect of room temperature superconductivity and beyond.

Superconductivity has therefore advanced to the state of a fountain of truth about the physical universe to an extent discoverers like Heike Kamerlingh Onnes, Walther

Meissner, and their students could not have dreamed of in their wildest fantasy. In a sense we have gone through a full scale from everyday items to the scope of the entire Universe. Ideas first introduced to understand the vanishing electrical resistance in a short length of metal wire, have generated conceptual developments relating to how we believe the whole Universe came to be and to function.

Most people will find such statements difficult to digest, yet fascinating. Physicists are also fascinated, but not equally surprised because we know that the whole game of Nature is based on a few laws that run The Big Show. And physics is about these laws. Fascination still remains, and will always be, a strong driving force in the human mind and spirit, related to the eternal question "where did it all begin," and "where are we going?" Superconductivity has turned out to be a great source of such fascination, and indeed also *useful* in every sense of the word.

I will dedicate the last paragraph of this book as a tribute to the memory of the wonderfully talented physicist and friend, Zlatko Tesanovic, by quoting from his letter, written just days before he died so tragically prematurely in the summer of 2012:

"Superconductivity is deeply embedded into the fabric of all modern physics, the much talked about Higgs mechanism and the eponymous particle being just another example of the Meissner effect in superconductors. Importantly, the phenomenon of superconductivity in all of its scientific facets holds considerable fascination for more general audiences. The story of original discovery in 1911, the decades-long quest for the microscopic theory, the spectacular discovery of high temperature superconductors in 1986 which revolutionized the field, all make a wonderful and exciting story which rivals any other great intellectual saga in modern science. Yet, compared to the deluge of God Particles, Elegant Universes, and Double Helixes of Life, it is one such saga that remains relatively unexplored in a more popular literature… This is history of science at its most intimate, told by those who made it. As the sun sets on the first century of superconductivity and its great heroes alike, their personal recollections and accounts will be both informative and entertaining, and will no doubt hold considerable value for the next generation of historians of science."

Index

A
A line of Shakespeare, 39
A. B. Pippard, 30
ABO_3, 108
Abrikosov lattice, 18
Academy of Science of the Soviet Union, 11
Adlai Stevenson, 75
Alex Malozemoff, 115, 121
Alex Müller, 106
Aluminium oxide, 60, 63
Amazing science, 134
Anderson, 36
Anderson and Rowell, 70
Anderson localisation, 74
Anderson murder mystery theorem, 85
Anderson-Higgs mechanism, 129
Atom bomb semester, 119
Autobiography, 16

B
BaLaCuO, 108
Bardeen, 33, 42, 45, 69
Battelle Memorial Institute, 118
BCM-theory, 37
BCS team, 46, 78
BCS theory, 5, 68
BCS-paper, 42
Bernd Matthias, 73, 77
Bessemer, 19
Bill Huntington, 59
Bill Little, 13, 99
Biographical notes, vii
Bohr and Mottelson, 49
Boundary between living and unliving, 38
Brian Josephson, 81
Brian Pippard, 69, 81
Bronx High School of Science, 31
Brown University, 47
Busch, 120

C
C. N. Yang and T. D. Lee, 32
Cambridge, 68
Canadian General Electric, 57
Cardiff, 67
Casimir, 24
Cellular machinery, 38
Centennial of superconductivity, vi
Ceramic materials, 117
Charlie Bean, 60
Charlie Slichter, 36
Childhood experiences, 133
Children's dreams, 19
Coherence factors, 47
Coherence length, 10
Coherent state, 33
Condensate field Ψ, 131
Consciousness, 39
Cooling devices, 2
Cooper pair tunnelling, 67
Cooper-pairs, 5
Crick and Watson, 48
Critical current, 23
Critical field, 23
Crystallography, 103

D
d-wave and p-wave superfluid, 73
d-wave in He3, 80
Dave Bohm, 78
David Pines, 78
de Gennes, 70
de Haas, 24
Debye frequency, 35

Debye temperature, 48
Decoration experiments, 27
Democratic Party, 75
Dimitri Abrikosov, 28
Dirty superconductor, 74
Dzyaloshinskii, 18

E
ECG, 116
École Normale, 92
Edison, 19
Electrical engineering, 44
Electron transport through a barrier, 54
Electronics physics, 76
Elias Burstein, 61
Eliashberg theory, 80
Endure hardship, 133
Energy gap, 60
Eric Courtens, 122
Essmann and Träuble, 27
Étienne Guyon, 95

F
Fascination, 134
Fermi sea, 33
Ferroelectricity, 77
Feynman, 5, 17, 26, 46
Flux-quantization, 69
Forerunners, v
Francis Low, 45
Francis Wheeler Loomis, 74
Fritz and Heinz London, 4
Fusion bomb, 27
Fusion process, 28

G
G. Busch, 118
Galileo, 38
Gauge invariance, 48, 73
Gauge invariant, 35
Gauge symmetry breaking, 43
General Electric, 54
General Electric course, 58
Gerd Binnig, 106
Giaever tunnelling, 6
Gilles Holst, 1
Ginzburg, 27, 123
Ginzburg-Landau, 97
Ginzburg-Landau equation, 25
Ginzburg-Landau theory, 5, 12, 22, 23
Global phase-stiffness, 129
God Particles, 135
Goldstone mode, 130
Goldstone theorem, 87
Gorkov, 18

Grand Écoles, 100
Grandes Écoles, 91
Granular matter, 90
Gravitational red-shift, 68
Green's functions, 26
Gregory Wannier, 77
Gutzwiller projection, 84
Guy Deutscher, 97, 121

H
H_{c2}, 25
Ham radio, 43
Hans Bueckner, 58
Hans Jörg Scheel, 105
Harry Thomas, 104, 122
Harvard, 74
Harwell, 68
Hebel and Slichter, 47
Heike Kamerlingh Onnes, v, 1
Heini Gränicher, 106
Heini Rohrer, 104
Helium, 2
Higgs boson, 5, 129
Higgs mechanism, vi, 43, 129
Higgs-generated mass, 131
High Temperature Superconductivity, 13
High-T_c, 51
Holy grail, 36
Hubbard model, 85
Hydrogen bomb, 133

I
IBM Thomas J. Watson Lab, 121
IBM Zurich Research Laboratory, 105
Idea of the order parameter, 14
Illinois, 36
Ilya Frank, 27
Inspiration, 133
Institute for Power Engineering, 20
Institute of Advanced Study at Princeton, 29
Ivar Giaever, 6, 71

J
Jahn-Teller, 99
Jahn-Teller polaron, 108, 120
Jim Phillips, 98
John Atkins, 71
John Bardeen, 29
John Fisher, 59
John Rowell, 82
Josephson, 37
Josephson effects, 6, 67
Josephson equations, 49

K
Kapitza, 10, 20, 28
kappa, 23
Karl Megerle, 61, 63
Kazan, 11
KGB, 24, 28
Khalatnikov, 18, 22
Kondo effect, 45
Kondo problem, 84
Korean wartime, 44

L
LaBaCuO, 111
lambda point, 15
Landau, 21, 26
Landau expansion, 10
Landau group, 4
Landau theory, 14
Lars Onsager, 46
Leggett, 16
Lenin, 19
Leo Esaki, 59
LHC facility, 115
Liquid crystals, 90, 96
Liquid helium, 15
London penetration depth λ, 4, 34
Los Alamos, 75

M
Mafia, 76
Manufacturing engineer, 56
March meeting, 35, 49
Masaaki Takashige, 110
Mathematics, 19
Matthew effect, 82
McMillan, 96
Meissner effect, 3
Meissner-Ochsenfeld effect, 4
Mendelssohn's sponge, 24
Michael Faraday, 19
Michael Tinkham, 122
MIT, 43
Mössbauer effect, 68
MRI, 126

N
Nambu, 87
National High Magnetic Field Laboratory (NHMFL), 42, 51
Naval Research Lab, 75, 76
Neutron diffraction, 27
Neutron stars, 134
Niels Bohr Institute, 42
NMR, 48
Non-perturbed interaction, 48
Normal state correlations, 51
Norwegian Institute of Technology, 53
Norwegian University of Science and Technology, 53
Nuclear bombs, 20
Nuclear group, 21

O
On the subway, 47
One-dimensional system, 13
Oppenheimer, 32, 77
Order parameter, 9
Orsay, 97
Oxford, 68

P
Pair correlations, 94
Pairing approximation, 46
Pairing condensate, 51
Pairing idea, 36
Pandora's box, 2
Parameter *kappa*, 15
Paul Chu, 112, 124
Paul Lauterburg, 60
Penicillin, 31
Perfect diamagnetism, 3
Perovskite materials, 104
Perovskites, 118
Persistent phenomenon, 4
Phase coherence, 85
Phase difference, 69
Phase transitions, 118
Phil Anderson, 68
Philippe Nozières, 79
Phonon mechanism, 48
Phonons in lead, 83
Physics of a Lifetime, 16
Physics Today, 70
Pierre Aigrain, 92
Pierre Morel, 79
Pinning, 25
Polaron model, 107
Popular lecturing, 100
Postwar Norway, 54, 57
Princeton, 78
Pseudospins, 68
psi-function, 9
psi-theory of superconductors, 13

R
Radio engineer, 119
Random phase approximation, 78
Raufoss Ammunitions Factory, 56
Raveau, 104, 108

Recognition, 35
Rennsselear Polytechnic Institute, 59
Renormalization group, 83
Resonance Valence Bond, 84
Risk, 107
Robert Ochsenfeld, 3
Robert Serber, 32
Rockets, 43
Rosetta stone, 19
Royal Society, 19
RVB-theory, 74

S
Santa Barbara, 51
Scaling, 90
Schafroth, 15
Schenectady, 58
Scherrer, 120
Schrieffer, 33
Schwinger, 74, 76
Second World War, 46
Self-made man, 58
Semiconductor surface physics, 44
Shubnikov, 24
Siemens, 126
Singularities, 25
Smolukovskii, 93
Soviet H-bomb, 9
Spin glass, 74
SQUID, 6
SQUID magnetometer, 110
$SrTiO_3$, 104, 109, 120
Stalin era, 9
Stalin Prize, 19
Standard Model, 130
Stripe domains, 119
Strong magnetic fields, 14
Success, 46
Superconducting transition temperature, T_c, 2
Superconductivity, 59
Superconductors of the second kind, 18
Superexchange, 79
Superfluid, 129
Superfluid helium, 94
Superfluidity, 12
Surface energy, 14, 23
Symmetry breaking, 134
"Symmetry-restoration" in extreme type-II, 131

T
Takashige, 124
Tallahassee, 51

Tamm, 11
Tanaka, 104, 124
TCNQ, 99
Ted Geballe, 77
Theoretical minimum, 17, 21
Thinking big, 45
Tinkham and Glover, 47
Tom Lehrer, 74, 76
Tomonaga, 47
Tomonaga approach, 42
Transformers, 126
Transistor, 36, 133
Trinity College, 82
Trondheim, 56
Truman, 75
Tsarist Russia, 10
Tunable Higgs-condensate, 130
Tunnelling, 6, 61
Tunnelling microscope, 65
Type II superconductor, 5, 18

U
$U(1)$-symmetry, 130
University of Illinois, 42, 44
University of Oslo, 55
University of Pennsylvania, 49
Unsuccessful experiments, 71

V
Very high T_c, 51
Vision, 64
Vitaly Ginzburg, 4
Vortex lattice, 6
Vortex liquids, 27
Vortex matter, 27

W
Wabash College, 74
Walter Kohn, 77
Walther Meissner, 3
Weak coupling approximation, 35
Weak link, 83
Westinghouse Science Talent, 31
Woodstock of physics, 7

Y
Yorktown Heights, 115

Z
Zavaritsky, 22
Zeitschrift für Physik, 123
Ziman, 68
Zlatko Tesanovic, 135

MIX
Papier aus verantwortungsvollen Quellen
Paper from responsible sources
FSC® C105338

If you have any concerns about our products,
you can contact us on
ProductSafety@springernature.com

In case Publisher is established outside the EU,
the EU authorized representative is:
**Springer Nature Customer Service Center GmbH
Europaplatz 3, 69115 Heidelberg, Germany**

Printed by Libri Plureos GmbH
in Hamburg, Germany